Holt Algebra 2

Homework and Practice Workbook
Teacher's Guide

HOLT, RINEHART AND WINSTON
A Harcourt Education Company
Orlando • Austin • New York • San Diego • Toronto • London

ISBN 0-03-078421-2
4 5 6 7 8 9 862 09 08 07

Contents

iii

Practice B

LESSON 1-1

Sets of Numbers

Order the given numbers from least to greatest. Then classify each number by the subsets of the real numbers to which it belongs.

1. $\frac{2}{3}$, $6.1\overline{7}$, $\sqrt{28}$, $-3\frac{1}{8}$, -4.9

   -4.9, $-3\frac{1}{8}$, $\frac{2}{3}$, $\sqrt{28}$, $6.1\overline{7}$; -4.9, $-3\frac{1}{8}$, $\frac{2}{3}$, and $6.1\overline{7}$ are rational numbers,

   and $\sqrt{28}$ is an irrational number.

2. 5π, $-6\sqrt{3}$, $\frac{-8}{3}$, $4.\overline{615}$, 0

   $-6\sqrt{3}$, $\frac{-8}{3}$, 0, $4.\overline{615}$, 5π; 0 is a whole number and an integer; 0, $\frac{-8}{3}$,

   and $4.\overline{615}$ are rational numbers; 5π and $-6\sqrt{3}$ are irrational numbers.

Rewrite each set in the indicated notation.

3. negative multiples of 3; set-builder notation

   $\{x \mid x = -3n \text{ and } n \text{ is a natural number}\}$

4. $[-4, 0)$ or $(10, 21)$; words

   All real numbers between

   -4 and 0 or between 10 and 21

   including -4 but excluding

   0, 10, and 21

5.

   roster notation

   $\{-3, 1, 5\}$

6.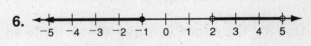

   set-builder notation

   $\{x \mid x \leq -1 \text{ or } 2 < x < 5\}$

The length of a necktie is generally in the range [52, 58] inches. The width of a tie is in the range [2.75, 3.5] inches. Use this information for Exercises 7–8.

7. Represent the range of the length of neckties in roster notation. Assume that all neckties come in whole-inch lengths.

   $\{52, 53, 54, 55, 56, 57, 58\}$

8. Represent the range of the width of neckties in set-builder notation.

   $\{x \mid 2.75 \leq x \leq 3.5\}$

Holt Algebra 2

Name _____ Date _____ Class _____

Practice B
Properties of Real Numbers

Find the additive and multiplicative inverse of each number.

1. -6

$$6; \frac{-1}{6}$$

2. $3\frac{1}{4}$

$$-3\frac{1}{4}, \frac{4}{13}$$

3. -0.7

$$0.7; \frac{-10}{7}$$

Identify the property demonstrated by each equation.

4. $x(a - b) = ax - bx$

Distributive

5. $m + (n + 6) = (n + 6) + m$

Commutative

6. $4(gh) = (4g)h$

Associative

7. $\frac{-\sqrt{5}}{w} \cdot \frac{w}{-\sqrt{5}} = 1$

Multiplicative Inverse

Use mental math to find each value.

8. 5% rebate on a $150 cell phone

$7.50

9. cost of 8 items at $12.98 each

$103.84

Classify each statement as sometimes, always, or never true. Give examples or properties to support your answer.

10. $d + (-d) = 0$

Always; Additive Inverse

11. $a + (bc) = (a + b) \cdot (a + c)$

Never;
$$1 + (2 \cdot 3) \neq (1 + 2) \cdot (1 + 3)$$

Use the table for Exercises 12–14. Write an expression to represent each total cost and then simplify it.

12. cost of 2 pens and 3 notebooks

$$2(\$2.89) + 3(\$1.79) = \$11.15$$

13. cost of 1 binder and 5 notebooks

$$5(\$1.79) + \$3.19 = \$12.14$$

School Supply Store	
Item	**Price**
Notebook	$1.79
Pen	$2.89
Binder	$3.19

14. cost of 3 notebooks at 20% discount, a binder at 25% discount, and 2 pens

$$3(1.79 - 0.36) + (3.19 - 0.80) + 2(2.89) = \$12.46$$

2

Holt Algebra 2

LESSON 1-3

Practice B
Square Roots

Estimate to the nearest tenth.

1. $\sqrt{78}$

8.8

2. $-\sqrt{57}$

−7.5

3. $\sqrt{39}$

6.2

Simplify each expression.

4. $\sqrt{243}$

$9\sqrt{3}$

5. $\dfrac{\sqrt{90}}{\sqrt{40}}$

$\dfrac{3}{2}$

6. $\sqrt{42} \cdot \sqrt{3}$

$3\sqrt{14}$

7. $-\dfrac{4}{\sqrt{144}}$

$-\dfrac{1}{3}$

8. $\sqrt{\dfrac{125}{5}}$

5

9. $-\sqrt{320}$

$-8\sqrt{5}$

Simplify by rationalizing each denominator.

10. $\dfrac{6}{\sqrt{5}}$

$\dfrac{6\sqrt{5}}{5}$

11. $\dfrac{-3\sqrt{15}}{\sqrt{3}}$

$-3\sqrt{5}$

12. $\dfrac{\sqrt{13}}{4\sqrt{6}}$

$\dfrac{\sqrt{78}}{24}$

Add or subtract.

13. $7\sqrt{5} - 10\sqrt{5}$

$-3\sqrt{5}$

14. $12\sqrt{3} + 3\sqrt{12}$

$18\sqrt{3}$

15. $-6\sqrt{50} + 4\sqrt{32}$

$-14\sqrt{2}$

Solve.

16. A building has a mural painted on an outside wall. The mural is a square with an area of 14,400 ft². What is the width of the mural?

120 ft

Holt Algebra 2

Name _____ Date _____ Class _____

Practice B
Simplifying Algebraic Expressions

Write an algebraic expression to represent each situation.

1. the measure of the complement of an angle with measure w ____$90 - w$____

2. the number of eggs in d cartons that each hold 1 dozen eggs ____$12d$____

Evaluate each expression for the given values of the variables.

3. $4t - 3s^2 + s^3$ for $t = -2$ and $s = -3$

____-62____

4. $\dfrac{5wp + 2w}{3wp^2}$ for $w = 4$ and $p = -1$

____-1____

Simplify each expression.

5. $-(4r - 3t) + 6r - t$

____$2r + 2t$____

6. $5(a + b) - 6(2a + 3b)$

____$-7a - 13b$____

Simplify each expression. Then evaluate the expression for the given values of the variables.

7. $-2(d - 3c) + 4d + c$
   for $d = 0$ and $c = -2$

____-14____

8. $-3f(2 - 3f + 4g) + g$
   for $f = -1$ and $g = 1$

____28____

Solve.

9. Marco delivers newspapers on the weekend. He delivers
   s newspapers on Saturday and $4s$ newspapers on Sunday.
   He earns $0.15 for each paper he delivers.

   a. Write an expression for the total amount of money
      Marco earns each weekend. ____$0.15(5s)$____

   b. Evaluate your expression for $s = 50$. ____$37.50____

   c. Write an expression for the amount of money Marco
      earns in a year if he delivers the same number of
      papers every weekend. ____$0.15(260s)$____

10. A tank holds 500 gallons of water. It starts out full, then
    10 gallons are released every minute.

    a. Write an expression for the number of gallons in the
       tank after m minutes. ____$500 - 10m$____

    b. Write an expression for the number of gallons in the tank
       after m minutes if 2 gallons are *also* added every minute. ____$500 - 8m$____

Holt Algebra 2

LESSON 1-5 Practice B
Properties of Exponents

Write each expression in expanded form.

1. $-3x^5$

2. $(j - 3k)^3$

$7 \cdot t \cdot t (-4r)(-4r)(-4r)(-4r)$

3. $7t^2(-4r)^4$

$-3 \cdot x \cdot x \cdot x \cdot x \cdot x$

$((j - 3k)j - 3k)(j - 3k)$

Evaluate each expression.

4. $-(-2)^{-4}$

5. $\left(\frac{5}{8}\right)^{-2}$

6. $\left(-\frac{3}{2}\right)^{-3}$

$\dfrac{-1}{16}$

$\dfrac{64}{25}$

$-\dfrac{8}{27}$

Simplify each expression. Assume all variables are nonzero.

7. $\dfrac{68f^5g^{-3}}{4f^{-3}g^6}$

$\dfrac{17f^8}{g^9}$

8. $(-4a^3b^7)^{-2}$

$\dfrac{1}{16a^6b^{14}}$

9. $6m^4n^9(-3m^2n^3)^{-2}$

$\dfrac{2n^3}{3}$

Evaluate each expression. Write the answer in scientific notation.

10. $(7.2 \times 10^{-5})(4.5 \times 10^3)$

11. $\dfrac{1.7 \times 10^5}{3.4 \times 10^9}$

12. $(7.8 \times 10^8)(2.8 \times 10^{11})$

3.24×10^{-1}

5.0×10^{-5}

2.184×10^{20}

Solve.

13. The A-1 Moving and Storage Company sells crates that measure x^2y units wide, x units long, and y^2 units tall. Find the volume of the crate.

x^3y^3 cubic units

14. The average lifespan for an adult living today is about 82 years. Some scientists believe that people born in the early part of this century may live up to 150 years. Calculate the number of minutes an 82-year-old and a 150-year-old could live. Round to the nearest million. Record the difference in scientific notation.

43 million; 79 million; 3.6×10^7

15. A movie made $\$6.7 \times 10^7$. It took 250 hours to film it. How much money was earned for each hour of filming? Write your answer in scientific notation.

2.68×10^5

Holt Algebra 2

Practice B

Relations and Functions

Give the domain and range for each relation. Then determine whether each relation is a function.

1.

Average High Temperatures

Month	Temperature
Jun	82°
Jul	88°
Aug	93°
Sep	82°

2.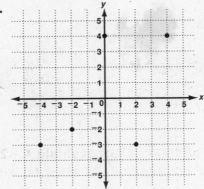

Domain: {Jun, Jul, Aug, Sep};

Range: {82°, 88°, 93°};
this is a function.

Domain: {−4, −2, 0, 2, 4};

Range: {−3, −2, 4};
this is a function.

Use the vertical-line test to determine whether each relation is a function. If not, identify two points a vertical line would pass through.

3.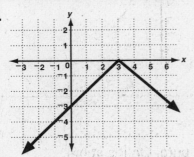

4.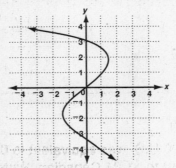

5.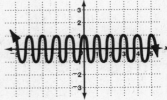

This is a function.

This is not a function;
(1, 1) (1, −4)

This is a function.

Explain whether each relation is a function.

6. {(1, 1), (2, 2), (3, 3), (4, 4)}

Yes, each value of *x* is associated with only 1 value of *y*.

7. from the model of car to the car's ID number

No, each car model is manufactured as many individual cars.

8. from the dates James took math tests to his test scores

Yes, there is only 1 score associated with each test date.

LESSON 1-7 Practice B
Function Notation

For each function, evaluate $f(-1)$, $f(0)$, $f\left(\frac{3}{2}\right)$.

1. $g(x) = -4x + 2$ **6, 2, −4**

2. $h(x) = x^2 - 3$ $-2, -3, -\frac{3}{4}$

3. $f(x) = 3x^2 + x$ $2, 0, 8\frac{1}{4}$

4. $f(x) = \frac{x}{2} - 1$ $-\frac{3}{2}, -1, -\frac{1}{4}$

Graph each function. Then evaluate $f(-2)$ **and** $f(0)$.

5. $f(x) = x^2 - 4$

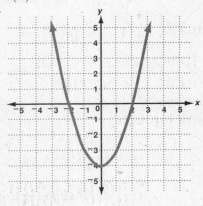

0, −4

6. $f(x) = -\frac{3}{2}x + 1$

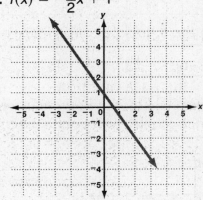

4, 1

Solve.

7. On one day the value of $1.00 U.S. was equivalent to 0.77 euro. On the same day $1.00 U.S. was equivalent to $1.24 Canadian. Write a function to represent the value of Canadian dollars in euros. What is the value of the function for an input of 5 rounded to the nearest cent, and what does it represent?

$$f(c) = \frac{0.77c}{1.24}; \ f(5) = 3.10;$$

the value of $5 Canadian is equivalent to 3.10 euros.

8. PC Haven sells computers at a 15% discount on the original price plus a $200 rebate. Write a function to represent the final price of a computer at PC Haven. What is the value of the function for an input of 2500, and what does it represent?

$$f(p) = 0.85p - 200 \ ; \ f(2500) = 1925; \ \$1925 \text{ is the final, discounted}$$

price of a computer with an original price of $2500.

Holt Algebra 2

LESSON 1-8 Practice B
Exploring Transformations

Perform the given translation on the point (2, 5) and give the coordinates of the translated point.

1. left 3 units

$$(-1, 5)$$

2. down 6 units

$$(2, -1)$$

3. right 4 units, up 2 units

$$(6, 7)$$

Use the table to perform each transformation of $y = f(x)$. Use the same coordinate plane as the original function.

4. translation left 1 unit, down 5 units

$x - 1$	x	y	$y - 5$
-4	-3	3	-2
-2	-1	1	-4
0	1	2	-3
1	2	1	-4
2	3	2	-3

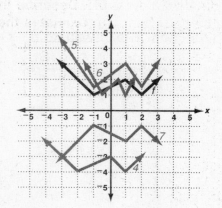

5. vertical stretch factor of $\frac{3}{2}$

x	y	$\frac{3}{2}y$
-3	3	$\frac{9}{2}$
-1	1	$\frac{3}{2}$
1	2	3
2	1	$\frac{3}{2}$
3	2	3

6. horizontal compression factor of $\frac{1}{2}$

$\frac{1}{2}x$	x	y
$-\frac{3}{2}$	-3	3
$-\frac{1}{2}$	-1	1
$\frac{1}{2}$	1	2
1	2	1
$\frac{3}{2}$	3	2

7. reflection across x-axis

x	y	$-y$
-3	3	-3
-1	1	-1
1	2	-2
2	1	-1
3	2	-2

Solve.

8. George has a goal for the number of computers he wants to sell each month for the next 6 months at his computer store. He draws a graph to show his projected profits for that period. Then he decides to discount the prices by 10%. How will this affect his profits? Identify the transformation to his graph and describe how to find the ordered pairs for the transformation.

$$\text{Profits are reduced by 10\%; vertical compression; } (x, 0.9y).$$

Holt Algebra 2

Name _____ Date _____ Class _____

Practice B
Introduction to Parent Functions

Identify the parent function for *h* from its function rule. Then graph *h* on your calculator and describe what transformation of the parent function it represents.

1. $h(x) = \sqrt{x + 4}$

Square root;
translation 4 units left

2. $h(x) = (x - 4)^3$

Cubic; translation
4 units right

3. $h(x) = 4x^2$

Quadratic; horizontal
compression

Graph the data from the table. Describe the parent function and the transformation that best approximates the data set.

4.

x	−2	−1	0	1	2
y	−9	−2	−1	0	7

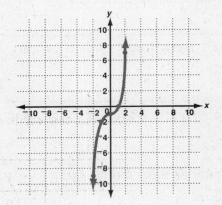

Cubic; translation 1 unit down

5.

x	0	2	8	18	32
y	0	1	2	3	4

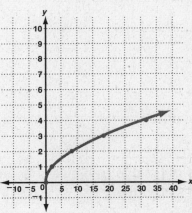

Square root; vertical compression

6. Compare the domain and the range for the parent quadratic function to the domain and the range for the parent linear function.

The domain is the same for both functions, all real numbers. The range

for the linear function is all real numbers, but the range for the quadratic

function is all real numbers greater than or equal to 0.

7. Compare the domain and the range for the parent square-root function to the domain and the range for the parent cubic function.

The domain and the range for the cubic function are all real numbers.

The domain and the range for the square-root function are all real

numbers greater than or equal to 0.

Holt Algebra 2

LESSON 2-1

Practice B
Solving Linear Equations and Inequalities

Solve.

1. $2(x - 3) = -4$

$x = 1$

2. $12 - 3(w + 7) = 15$

$w = -8$

3. $4(8 - p) - (7 - p) = 22$

$p = 1$

4. $18 - 4y = -2(6 + 2y)$

$y = \varnothing$

5. $7t + 6 - 2\left(5 + \dfrac{3t}{2}\right) = 5t - 11$

$t = 7$

6. $32 + 4(c - 1) = -(4c + 5)$

$c = -4\dfrac{1}{8}$

Solve and graph.

7. $-5x + 7 \geq -3$

$x \leq 2$

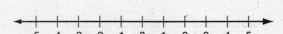

8. $4 - (-7 - k) > 2(k + 3)$

$k < 5$

9. $-18d + 5(8 + 3d) \leq 7(3d - 8)$

$d \geq 4$

Solve.

10. Yvonne's cell phone plan gives her a maximum of 200 minutes each month.

    a. Suppose Yvonne's calls average 7 minutes. What is the maximum number of calls she can make each month?

        28

    b. Yvonne knows she has used 61 minutes during the first week of this month. If she limits her calls to 15 per week for the remaining 3 weeks this month, what is the maximum length of time rounded to the nearest minute that she can use for each call?

        3 minutes

11. Blair wants to spend less than $50 at the grocery store. He already has $37 worth of groceries in his shopping cart and is going to buy some fresh vegetables for $0.75 each. What numbers of vegetables v can he buy and stay under his spending limit?

        $v \leq 17$

Holt Algebra 2

Name _____ Date _____ Class _____

Practice B
Proportional Reasoning

Solve each proportion.

1. $\dfrac{28}{36} = \dfrac{g}{81}$

$g = 63$

2. $\dfrac{z}{1.75} = \dfrac{64}{21}$

$z = 5\dfrac{1}{3}$

3. $\dfrac{3}{0.6} = \dfrac{1.05}{n}$

$h = 0.21$

4. $\dfrac{5}{8} = \dfrac{f - 1}{56}$

$f = 36$

5. $\dfrac{2.4}{1.8} = \dfrac{0.004}{y}$

$y = 0.003$

6. $\dfrac{5}{v + 6} = \dfrac{4}{12}$

$v = 9$

Solve.

7. $\triangle XYZ$ has vertices $X(0, 0)$, $Y(0, 10)$, and $Z(-10, 10)$. $\triangle XWT$ is similar to $\triangle XYZ$ and has a vertex at $W(0, 4)$. Graph $\triangle XYZ$ and $\triangle XWT$ on the same grid.

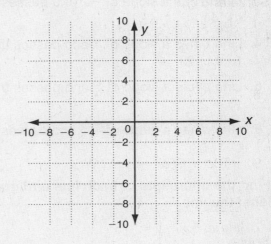

8. Dan works as a house painter. He knows that it is safe to place the base of his 10-foot ladder 3 feet from the base of a house. Today he has to use a 25-foot ladder. Dan wants to keep the same ratios in order to be safe. How far should Dan place the base of his 25-foot ladder from the base of the house?

7.5 ft

9. The school newspaper took a survey. Of the students polled, 15% said they did not have too much homework. Sixty students were polled for the survey. How many students said they did not have too much homework?

9 students

10. Cheryl wants to measure the distance across a stream. She took some measurements and drew a diagram. How wide is the stream?

$22\dfrac{2}{3}$ yd

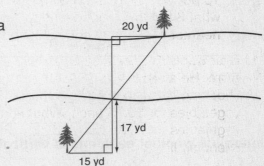

Holt Algebra 2

LESSON 2-3

Practice B

Graphing Linear Functions

Determine whether each data set could represent a linear function.

1.

x	9	7	5	3
f(x)	2	5	10	15

2.

x	0.5	1	1.5	2
f(x)	9	6	3	0

_____ **Nonlinear** _____

_____ **Linear** _____

Use the coordinate plane at right to graph and label each line.

3. Line *a* has a slope of −2 and passes through (1, 4).

4. Line *b* has a slope of 1 and passes through (−4, −2).

5. Line *c* has a slope of $\frac{2}{3}$ and passes through (3, −2).

6. Line *d* has a slope of $\frac{-5}{4}$ and passes through (−1, 0).

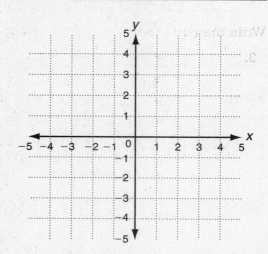

Find the intercepts of each line and graph and label the line.

7. line *e*: $5x + y = -5$

   ___ ***x*-intercept = −1; *y*-intercept = −5** ___

8. line *f*: $6x + 2y = 6$

   ___ ***x*-intercept = 1; *y*-intercept = 3** ___

Write each function in slope-intercept form. Then graph and label the function.

9. line *g*: $-3x - y = 9$

   ___ **$y = -3x - 9$** ___

10. line *h*: $4x + 3y = 6$

   ___ **$y = \frac{-4x}{3} + 2$** ___

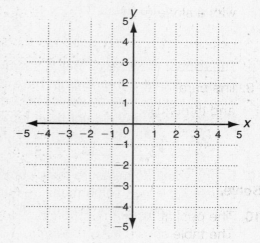

Determine whether each line is vertical or horizontal.

11. $x = -5$

   _____ **Vertical** _____

12. $y = \frac{8}{3}$

   _____ **Horizontal** _____

13. $x = 4.6$

   _____ **Vertical** _____

Holt Algebra 2

Name _____ Date _____ Class _____

LESSON 2-4

Practice B
Writing Linear Functions

Find the slope of each line.

1.

x	−5	1	4	9
y	−9	3	9	19

2

2.

x	−7	−2	6	13
y	−0.5	2	6	9.5

$\dfrac{1}{2}$

Write the equation of each line in slope-intercept form.

3.

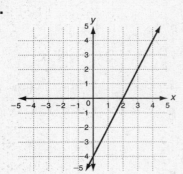

$y = 2x - 4$

4.

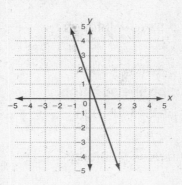

$y = -3x + 1$

5.

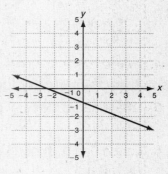

$y = -\dfrac{2}{5}x - 1$

6. line passing through $(-3, -4)$ with a slope of $\dfrac{1}{5}$

$$y = \dfrac{1}{5}x - \dfrac{17}{5}$$

7.

x	−2	3	8	11
y	−1	1.5	4	5.5

$$y = \dfrac{1}{2}x$$

8. line parallel to $y = -\dfrac{3}{2}x + 4$ and through $(1, 5)$

$$y = -\dfrac{3}{2}x + \dfrac{13}{2}$$

9. line perpendicular to $y = -2x + 11$ and through $(4, -2)$

$$y = \dfrac{1}{2}x - 4$$

Solve.

10. The pool at the Barnes Community Center is heated. The table shows the temperature of the water at various time intervals after the heater is turned on.

   a. Express the temperature of the water as a function of time.

$$T = 2h + 56$$

Swimming Pool Heater	
Time (h)	**Temperature (*T*)**
0	56°F
3	62°F
5	66°F
9	74°F

   b. Find the temperature of the water after 12 hours.

80°F

Holt Algebra 2

Practice B
Linear Inequalities in Two Variables

Graph each inequality.

1. $y < x + 2$

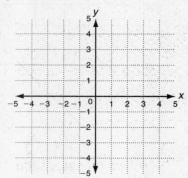

2. $y \geq 3x - 5$

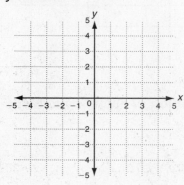

Solve each inequality for *y*. Graph the solution.

3. $-2(3x + 2y - 3) \geq 12$

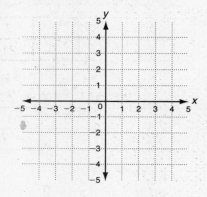

4. $\dfrac{-x}{5} + \dfrac{2y}{3} > 0$

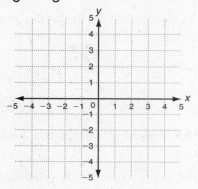

Solve.

5. Marcus volunteers to work at a carnival booth selling raffle tickets. The tickets cost $2 each or 3 for $5. His goal is to have at least $250 in sales during his shift.

a. Let *x* be the number of tickets sold for $2 each. Let *y* be the number of tickets sold in sets of 3 for $5. Write and graph an inequality for the total number of tickets Marcus must sell to meet his goal.

$$2x + \frac{5y}{3} \geq 250$$

b. If Marcus sells 75 tickets for $2 each, what is the least number of tickets he must sell in sets of 3 to meet his goal?

60 tickets

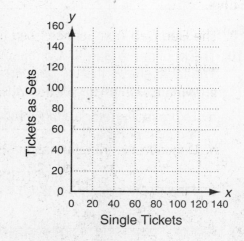

Holt Algebra 2

LESSON 2-6

Practice B
Transforming Linear Functions

Let $g(x)$ be the indicated transformation of $f(x)$. Write the rule for $g(x)$.

1.

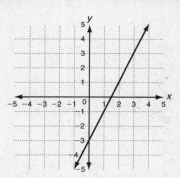

horizontal translation
left 3 units

$$g(x) = 2x + 3$$

2.

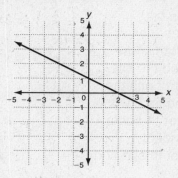

vertical compression by
a factor of $\frac{1}{5}$

$$g(x) = -\frac{1}{10}x + \frac{1}{5}$$

3.

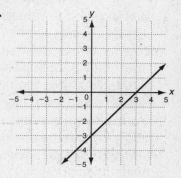

reflection across the
y-axis

$$g(x) = -x - 3$$

4. linear function defined by the table; horizontal stretch by
a factor of 2.3

$$g(x) = 4.6x + 7$$

x	−5	0	7
y	−3	7	21

5. $f(x) = 1.7x - 3$; vertical compression by a factor of 0.7 $g(x) = 1.19x - 2.1$

**Let $g(x)$ be the indicated combined transformation of $f(x) = x$. Write
the rule for $g(x)$.**

6. vertical translation down 2 units followed by a
horizontal compression by a factor of $\frac{2}{5}$ $$g(x) = \frac{5}{2}x - 2$$

7. horizontal stretch by a factor of 3.2 followed by
a horizontal translation right 3 units $$g(x) = 3.2x - 9.6$$

Solve.

8. The Red Cab Taxi Service used to charge $1.00 for the first $\frac{1}{5}$ mile and $0.75 for each

additional $\frac{1}{5}$ mile. The company just raised its rates by a factor of 1.5.

a. Write a new price function $g(x)$ for a taxi ride.

$$g(x) = 1.5[1 + 0.75\,(5x - 1)] = 5.625x + 0.375$$

b. Describe the transformation(s) that have been applied.

Vertical stretch by a factor of 1.5

Holt Algebra 2

Name _____ Date _____ Class _____

Practice B
Curve Fitting with Linear Models

Solve.

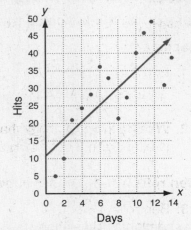

1. Vern created a website about his school's sports teams. He has a hit counter on his site that lets him know how many people have visited the site. The table shows the number of hits the site received each day for the first two weeks. Make a scatter plot for the data using the day as the independent variable. Sketch a line of best fit and find its equation.

Lincoln High Website													
Day	1	2	3	4	5	6	7	8	9	10	11	12	13	14
Hits	5	10	21	24	28	36	33	21	27	40	46	50	31	38

Possible answer: $y = 2.5x + 11$

2. A photographer hiked through the Grand Canyon. Each day she filled a photo memory card with images. When she returned from the trip, she deleted some photos, saving only the best. The table shows the number of photos she kept from all those taken for each memory card.

Grand Canyon Photos	
Photos Taken	**Photos Kept**
117	25
128	31
140	39
157	52
110	21
188	45
170	42

a. Use a graphing calculator to make a scatter plot of the data. Use the number of photos taken as the independent variable.

b. Find the correlation coefficient.

$r = 0.848$

c. Write the equation of the line of best fit.

$y = 0.33x - 11.33$

d. Predict the number of photos this photographer will keep if she takes 200 photos.

Possible answer: about 50 photos

3. What is the relationship between the slope of a line and its correlation coefficient?

Possible answer: If the slope is negative, the correlation coefficient is negative. If the slope is positive, the correlation coefficient is positive.

Holt Algebra 2

Name _____ Date _____ Class _____

Practice B
Solving Absolute-Value Equations and Inequalities

Solve each equation.

1. $|2x + 1| = 7$

   $x = 3 \text{ or } x = -4$

2. $|-7x| = 28$

   $x = \pm 4$

3. $3|3x| - 7 = 2$

   $x = \pm 1$

4. $|2x - 5| = 5$

   $x = 0 \text{ or } x = 5$

5. $2|x + 1| = 14$

   $x = 6 \text{ or } x = -8$

6. $|4 - x| + 2 = 9$

   $x = -3 \text{ or } x = 11$

Solve each inequality or compound inequality. Then graph the solution.

7. $-4x + 2 > -10$ and $5x - 12 < 8$

   $x < 4$

   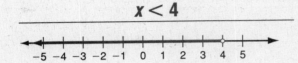

8. $3x - 4 \geq 8$ or $-x + 12 > 16$

   $x \geq 4 \text{ or } x < -4$

   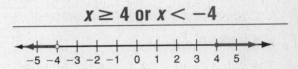

9. $|9x| \geq 18$

   $x \leq -2 \text{ or } x \geq 2$

   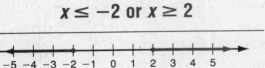

10. $|3x - 7| > 8$

    $x < -\dfrac{1}{3} \text{ or } x > 5$

    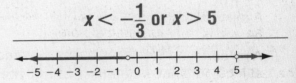

11. $|0.3x| > 1$

    $x < -\dfrac{10}{3} \text{ or } x > \dfrac{10}{3}$

    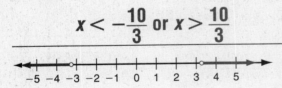

12. $|7x| - 12 \leq 9$

    $x \geq -3 \text{ and } x \leq 3$

    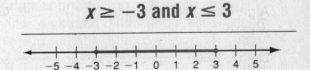

Solve.

13. Any measurement is accurate within ± 0.5 of the measurement unit. For example, if you measure your pencil to the nearest inch, your measurement could be 0.5 inch too long or 0.5 inch too short. Write an absolute-value inequality that shows the maximum and minimum actual measure of a nail measured to be 4.4 centimeters to the nearest 0.1 centimeter.

    $|m - 4.4| \leq 0.05$

Holt Algebra 2

LESSON 2-9

Practice B
Absolute-Value Functions

Perform each transformation on $f(x) = |2x| + 3$. Write the transformed function $g(x)$.

1. down 7 units

2. reflect across y-axis

3. left 5 units

$g(x) = |2x| - 4$

$g(x) = |-2x| + 3$

$g(x) = |2x + 5| + 3$

Translate $f(x) = |x|$ so that the vertex is at the given point.

4. $(6, -3)$

5. $(-8, -1)$

6. $(-7, 2)$

$f(x) = |x - 6| - 3$

$f(x) = |x + 8| - 1$

$f(x) = |x + 7| + 2$

Perform the transformation. Then graph.

7. Compress $f(x) = |3x - 4|$ vertically by a factor of $\frac{1}{3}$.

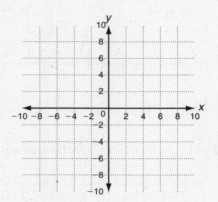

$$f(x) = \frac{|3x - 4|}{3}$$

Solve.

8. At a sugar plantation processing plant, a machine fills a bag with 5 pounds of white sugar. For quality control, another machine weighs each bag in ounces and rejects bags that differ from 5 pounds by more than y ounces.

a. Write an absolute-value function to show the minimum and maximum weight of sugar that could be in each bag.

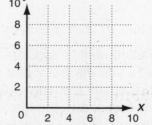

$$y = |x - 5|$$

b. Graph the function.

c. Describe the transformation from $f(x) = |x|$.

Translation 5 units to the right

Holt Algebra 2

Practice B
LESSON 3-1
Using Graphs and Tables to Solve Linear Systems

Classify each system, and determine the number of solutions.

1. $\begin{cases} y = -4x + 7 \\ 12x + 3y = 21 \end{cases}$

2. $\begin{cases} 5y = x - 10 \\ y = \frac{x}{5} + 3 \end{cases}$

3. $\begin{cases} x + 6y = -2 \\ 12x - 6y = 0 \end{cases}$

Consistent, dependent; infinitely many solutions

Inconsistent; no solutions

Consistent, independent; one solution

Use substitution to determine if the given ordered pair is an element of the solution set for the system of equations. If it is not, give the correct solution.

4. $(-4, 8)$ $\begin{cases} y = -2x \\ 3x + y = -4 \end{cases}$ **It is the solution.**

5. $(11, 3)$ $\begin{cases} y = x - 8 \\ x + 4y = -2 \end{cases}$ **(6, −2)**

6. $(4, 1)$ $\begin{cases} y = 5x - 1 \\ 8 = 4x + y \end{cases}$ **(1, 4)**

7. $(5, -5)$ $\begin{cases} x + y = 10 \\ x - y = 0 \end{cases}$ **(5, 5)**

8. $(2, -1)$ $\begin{cases} 2x + 3y = -8 \\ 3x - 4y = 5 \end{cases}$ **(−1, −2)**

9. $(0, 3)$ $\begin{cases} 3x + 5y = 15 \\ x - y = -3 \end{cases}$ **It is the solution.**

Solve by graphing a system of equations.

10. A puppy pen is 1 foot longer than twice its width. John wants to increase the length and width by 5 feet each to enlarge the area by 90 square feet. What will be the area of the new pen?

126 square feet

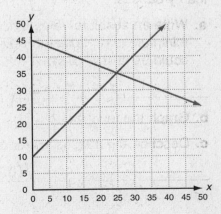

11. Keesha has 10 more quarters than dimes, which, together, total $11.25. How many coins does she have in quarters and dimes?

35 quarters + 25 dimes = 60 coins

Holt Algebra 2

Practice B

LESSON 3-2

Using Algebraic Methods to Solve Linear Systems

Use substitution to solve each system of equations.

1. $\begin{cases} x = 7y - 4 \\ 2x - 3y = 14 \end{cases}$

  _____ $(10, 2)$ _____

2. $\begin{cases} y - 3x = 5 \\ 2x = 3y + 6 \end{cases}$

  _____ $(-3, -4)$ _____

3. $\begin{cases} 3x - 4y = 20 \\ y - 2x = 0 \end{cases}$

  _____ $(-4, -8)$ _____

Use elimination to solve each system of equations.

4. $\begin{cases} x + 6y = 1 \\ 3x + 5y = -10 \end{cases}$

  _____ $(-5, 1)$ _____

5. $\begin{cases} 3x + 4y = 6 \\ 2x + 3y = 3 \end{cases}$

  _____ $(6, -3)$ _____

6. $\begin{cases} 3x - 5y = 1 \\ 2x + 3y = -12 \end{cases}$

  _____ $(-3, -2)$ _____

Use substitution or elimination to solve each system of equations.

7. $\begin{cases} x + y = 13 \\ 2x - 3y = 1 \end{cases}$

  _____ $(8, 5)$ _____

8. $\begin{cases} 9x + 2y = 5 \\ 3x - y = -10 \end{cases}$

  _____ $(-1, 7)$ _____

9. $\begin{cases} 2x + y = 1 \\ x = 5 + y \end{cases}$

  _____ $(2, -3)$ _____

10. $\begin{cases} x = -8y \\ x + y = 14 \end{cases}$

  _____ $(16, -2)$ _____

11. $\begin{cases} 2x + 4y = 12 \\ -3x + 3y = 63 \end{cases}$

  _____ $(-12, 9)$ _____

12. $\begin{cases} 5x - 2y = -1 \\ 3x - y = -2 \end{cases}$

  _____ $(-3, -7)$ _____

Solve.

13. Bill leaves his house for Makayla's house riding his bicycle at 8 miles per hour. At the same time, Makayla leaves her house heading toward Bill's house walking at 3 miles per hour.

   a. Write a system of equations to represent the distance, *d,* each is from Makayla's house in *h* hours. They live 8.25 miles apart.

$$\begin{cases} d = 8.25 - 8h \\ d = 3h \end{cases}$$

   b. Solve the system to determine how long they travel before meeting.

   _____ **0.75 h or 45 min** _____

Holt Algebra 2

LESSON 3-3 Practice B
Solving Systems of Linear Inequalities

Graph each system of inequalities.

1. $\begin{cases} y \le 3x - 5 \\ y < -\frac{1}{2}x + 4 \end{cases}$

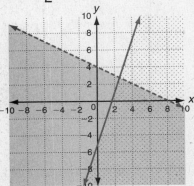

2. $\begin{cases} y < x + 5 \\ y \ge 4x - 2 \end{cases}$

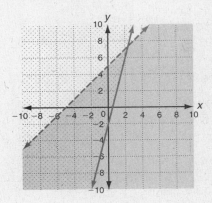

Graph the system of inequalities, and classify the figure created by the solution region.

3. $\begin{cases} x \le 2 \\ x \ge -3 \\ y \le 2x + 2 \\ y \ge 2x - 1 \end{cases}$ <u>**Parallelogram**</u>

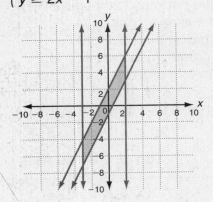

4. $\begin{cases} y \le -x + 4 \\ y \le 3 \\ y \ge 0 \\ y \ge -2x - 1 \end{cases}$ <u>**Trapezoid**</u>

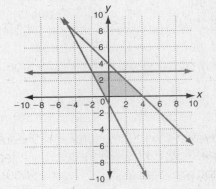

Solve.

5. The Thespian Club is selling tickets to its annual variety show. Prices are $8 for an adult ticket and $4 for a student ticket. The club needs to raise $1000 to pay for costumes and stage sets. The auditorium has a seating capacity of 240. Write and graph a system of inequalities that can be used to determine how many tickets have to be sold for the club to meet its goal.

$\begin{cases} 8x + 4y \ge 1000 \\ x + y \le 240 \end{cases}$

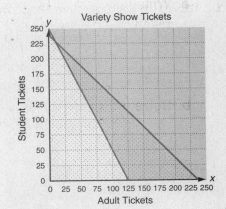

Variety Show Tickets

Holt Algebra 2

Name _____ Date _____ Class _____

Practice B
Linear Programming

Maximize or minimize each objective function.

1. Maximize $P = 5x + 2y$

    for the constraints $\begin{cases} y \geq 0 \\ x \geq 0 \\ y \leq -x + 10 \\ y \leq 2x + 1 \end{cases}$

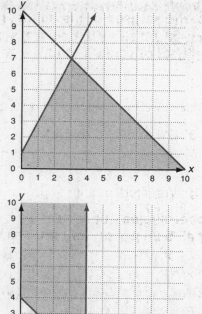

    _____ **(10, 0)** _____

2. Minimize $P = 4x + 6y$

    for the constraints $\begin{cases} 0 \leq x \leq 4 \\ y \geq 1 \\ y \geq -x + 4 \end{cases}$

    _____ **(3, 1)** _____

Solve.

3. A grocer buys cases of almonds and walnuts. Almonds are packaged 20 bags per case. The grocer pays $30 per case of almonds and makes a profit of $17 per case. Walnuts are packaged 24 bags per case. The grocer pays $26 per case of walnuts and makes a profit of $15 per case. He orders no more than 300 bags of almonds and walnuts together at a maximum cost of $400.

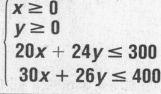

$\begin{cases} x \geq 0 \\ y \geq 0 \\ 20x + 24y \leq 300 \\ 30x + 26y \leq 400 \end{cases}$

   a. Write the constraints. Use x for the number of cases of almonds ordered and y for the number of cases of walnuts ordered.

   b. Graph the constraints.

   c. Write the objective function for the profit.

    _____ $P = 17x + 15y$ _____

   d. How many cases of almonds and walnuts maximize the grocer's profit?

    9 cases of almonds, 5 cases of walnuts

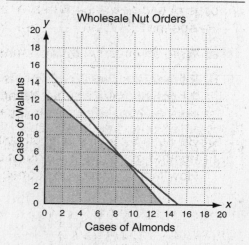

Wholesale Nut Orders

Cases of Walnuts

Cases of Almonds

Holt Algebra 2

LESSON 3-5 Practice B
Linear Equations in Three Dimensions

Graph each linear equation in three-dimensional space.

1. $8x + 16y + 4z = 16$

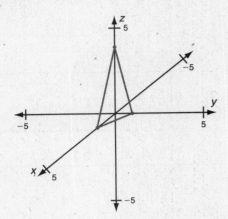

2. $-6x + 8y - 12z = 24$

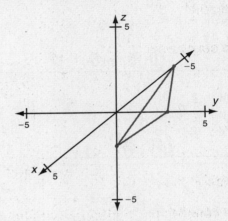

3. $4x + 3y + 6z = -12$

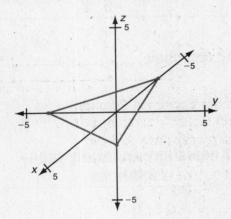

4. $10x + 15y - 6z = -30$

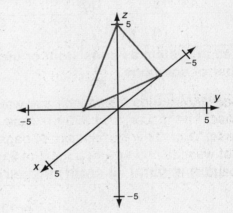

Solve.

5. Bill is buying bulbs for his flower garden. Bags of iris bulbs are $4 each, bags of tulip bulbs are $3 each, and bags of daffodil bulbs are $2 each. He spends $24 in all.

   a. Write an equation to represent the situation. $4x + 3y + 2z = 24$

   b. Bill wants to buy 3 bags of iris bulbs and at least 2 bags of daffodil bulbs. What is the maximum number of bags of tulip bulbs he can buy?   **2 bags**

   c. Bill buys 5 bags of daffodil bulbs and the same number of bags of tulip bulbs as iris bulbs. How many bags of each does he buy?   **2 bags of each**

Holt Algebra 2

LESSON 3-6 Practice B
Solving Linear Systems in Three Variables

Use elimination to solve each system of equations.

1. $\begin{cases} x + y - 2z = 10 \\ 8x - 9y - z = 5 \\ 3x + 4y + 2z = -10 \end{cases}$

$$(0, 0, -5)$$

2. $\begin{cases} 6x + 3y + 4z = 3 \\ x + 2y + z = 3 \\ 2x - y + 2z = 1 \end{cases}$

$$(-2, 1, 3)$$

3. $\begin{cases} x + y + z = 0 \\ x - y + z = 14 \\ x - y - z = 16 \end{cases}$

$$(8, -7, -1)$$

4. $\begin{cases} 8x + 3y - 6z = 4 \\ x - 2y - z = 2 \\ 4x + y - 2z = -4 \end{cases}$

$$(-4, 0, -6)$$

5. $\begin{cases} 2x - y - z = 1 \\ 3x + 2y + 2z = 12 \\ x - y + z = 9 \end{cases}$

$$(2, -2, 5)$$

6. $\begin{cases} 2x - y + 3z = 7 \\ 5x - 4y - 2z = 3 \\ 3x + 3y + 2z = -8 \end{cases}$

$$(-1, -3, 2)$$

Classify each system as consistent or inconsistent, and determine the number of solutions.

7. $\begin{cases} 2x - 6y + 4z = 3 \\ -3x + 9y - 6z = -3 \\ 5x - 15y + 10z = 5 \end{cases}$

Inconsistent; 0 solutions

8. $\begin{cases} -4x + 2y + 2z = -2 \\ 2x - y - z = 1 \\ x + y + z = 2 \end{cases}$

Consistent; infinitely many solutions

Solve.

9. At the arcade Sami won 2 blue tickets, 1 yellow ticket and 3 red tickets for 1500 total points. Jamal won 1 blue ticket, 2 yellow tickets, and 2 red tickets for 1225 total points. Yvonne won 2 blue tickets, 3 yellow tickets, and 1 red ticket for 1200 total points Write and solve a system of equations to determine the point value of each type of ticket.

$$\begin{cases} 2b + y + 3r = 1500 \\ b + 2y + 2r = 1225 \\ 2b + 3y + r = 1200 \end{cases}$$

blue tickets: 125 points; yellow tickets: 200 points; red tickets: 350 points

Holt Algebra 2

Name _____ Date _____ Class _____

Practice B
Matrices and Data

The table shows the prices for various passes to a theme park.

Theme Park Pass Price List				
Type of Pass	1-Day	2-Day	3-Day	5-Day
Basic	$40	$75	$100	$125
Super	$70	$95	$120	$140
Deluxe	$80	$105	$130	$150

1. Display the data in the form of a matrix, P.

$$P = \begin{bmatrix} 40 & 75 & 100 & 125 \\ 70 & 95 & 120 & 140 \\ 80 & 105 & 130 & 150 \end{bmatrix}$$

2. What are the dimensions of P?

$$3 \times 4$$

3. What is the entry at p_{31} and what does it represent?

$80; the cost of a 1-day deluxe pass

4. What is the address of the entry $120? p_{23}

5. Write an expression that represents a matrix that shows the cost of buying theme park tickets for a family of four. $4P$

Use the following matrices for Exercises 6–8. Add or subtract, if possible.

$$R = \begin{bmatrix} 4 & 12 \\ 0 & -6 \\ 9 & 15 \end{bmatrix} \qquad S = \begin{bmatrix} -3 & 9 & 2 \\ 10 & -5 & 4 \\ 1 & 2 & 3 \end{bmatrix} \qquad T = \begin{bmatrix} -5 & 9 \\ -3 & 6 \\ 10 & 5 \end{bmatrix}$$

6. $S - R$ 7. $T + R$ 8. $R - T$

Not possible $\begin{bmatrix} -1 & 21 \\ -3 & 0 \\ 19 & 20 \end{bmatrix}$ $\begin{bmatrix} 9 & 3 \\ 3 & -12 \\ -1 & 10 \end{bmatrix}$

Use the following matrices for Exercises 9–11. Evaluate, if possible.

$$X = \begin{bmatrix} 5 & -2 & 0 & 9 \\ 4 & 16 & -5 & 6 \end{bmatrix} \qquad Y = \begin{bmatrix} -6 & 4 & 10 & 8 \\ 13 & 6 & 0 & -2 \end{bmatrix}$$

9. $3X$ 10. $3Y + 4X$ 11. $3X - 2Y$

$\begin{bmatrix} 15 & -6 & 0 & 27 \\ 12 & 48 & -15 & 18 \end{bmatrix}$ $\begin{bmatrix} 2 & 4 & 30 & 60 \\ 55 & 82 & -20 & 18 \end{bmatrix}$ $\begin{bmatrix} 27 & -14 & -20 & 11 \\ -14 & 36 & -15 & 22 \end{bmatrix}$

Solve.

12. If $E = \begin{bmatrix} 2 & -3 \\ 5 & 4 \end{bmatrix}$ and $E + F = \begin{bmatrix} 0 & 0 \\ 0 & 0 \end{bmatrix}$, find F. $F = \begin{bmatrix} -2 & 3 \\ -5 & -4 \end{bmatrix}$

Holt Algebra 2

LESSON 4-2 Practice B
Multiplying Matrices

Tell whether each product is defined. If so, give its dimensions.

1. $P_{3 \times 3}$ and $Q_{3 \times 4}$; PQ

$$3 \times 4$$

2. $R_{3 \times 8}$ and $S_{4 \times 3}$; SR

$$4 \times 8$$

3. $W_{2 \times 5}$ and $X_{2 \times 5}$; WX

No

Use the following matrices for Exercises 4–7. Evaluate, if possible.

$$E = \begin{bmatrix} -4 & 1 \\ -2 & 2 \end{bmatrix} \quad F = \begin{bmatrix} 1 & 0 \\ 4 & -3 \\ -2 & 6 \\ -1 & 5 \end{bmatrix} \quad G = \begin{bmatrix} -4 & 0 & 3 & 5 \\ 1 & -2 & 0 & 0 \end{bmatrix} \quad H = \begin{bmatrix} 1 & -2 & -1 & 3 \\ 2 & 0 & 4 & -1 \\ 3 & 5 & -2 & 2 \\ 1 & -1 & 0 & 0 \end{bmatrix}$$

4. EG

$$\begin{bmatrix} 17 & -2 & -12 & -20 \\ 10 & -4 & -6 & -10 \end{bmatrix}$$

5. HF

$$\begin{bmatrix} -8 & 15 \\ -5 & 19 \\ 25 & -17 \\ -3 & 3 \end{bmatrix}$$

6. FG

$$\begin{bmatrix} -4 & 0 & 3 & 5 \\ -19 & 6 & 12 & 20 \\ 14 & -12 & -6 & -10 \\ 9 & -10 & -3 & -5 \end{bmatrix}$$

7. E^2

$$\begin{bmatrix} 14 & -2 \\ 4 & 2 \end{bmatrix}$$

Solve.

8. Jamal, Ken, and Barry are playing a baseball video game. The first table shows the number of singles, doubles, triples, and home runs each scored. Find the total number of points they each scored.

Hits				
Player	**S**	**D**	**T**	**HR**
Jamal	3	2	0	1
Ken	2	4	0	0
Barry	0	1	3	1

a. Write a matrix that represents the data in each table.

$$\begin{bmatrix} 3 & 2 & 0 & 1 \\ 2 & 4 & 0 & 0 \\ 0 & 1 & 3 & 1 \end{bmatrix}, \begin{bmatrix} 1 \\ 2 \\ 3 \\ 4 \end{bmatrix}$$

Points Scored for Hits	
Hit	**Points**
Single (S)	1
Double (D)	2
Triple (T)	3
Home run (HR)	4

b. Find the product matrix.

$$\begin{bmatrix} 11 \\ 10 \\ 15 \end{bmatrix}$$

c. How many points did each player score?

Jamal 11, Ken 10, Barry 15

Holt Algebra 2

Name _____ Date _____ Class _____

LESSON
4-3
Practice B
Using Matrices to Transform Geometric Figures

Triangle *JKL* has vertices *J*(−3, 1), *K*(2, 2), and *L*(1, −2).

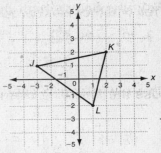

Use a matrix to transform triangle *JKL*. Find the coordinates of the vertices of the image.

1. Translate 5 units right, 6 units down.

$$J'(2, -5),\ K'(7, -4),\ L'(6, -8)$$

2. Translate 2 units left, 4 units up.

$$J'(-5, 5),\ K'(0, 6),\ L'(-1, 2)$$

3. Enlarge by a factor of 7.

$$J'(-21, 7),\ K'(14, 14),$$
$$L'(7, -14)$$

4. Reduce by a factor of 0.25.

$$J'(-0.75, 0.25),\ K'(0.5, 0.5),$$
$$L'(0.25, -0.5)$$

Reflect or rotate triangle *ABC* with vertices *A*(−2, 1), *B*(−1, 4), and *C*(2, 2). Find the coordinates of the vertices of the image. Describe the transformation.

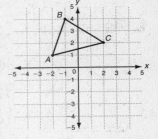

5. $\begin{bmatrix} -1 & 0 \\ 0 & 1 \end{bmatrix}$

$$A'(2, 1),\ B'(1, 4),\ C'(-2, 2);$$
reflection across the *y*-axis

6. $\begin{bmatrix} 0 & 1 \\ -1 & 0 \end{bmatrix}$

$$A'(1, 2),\ B'(4, 1),\ C'(2, -2);$$
90° clockwise rotation

7. $\begin{bmatrix} 0 & -1 \\ 1 & 0 \end{bmatrix}$

$$A'(-1, -2),\ B'(-4, -1),$$
$$C'(-2, 2);\ \text{90° counterclockwise}$$
rotation

8. $\begin{bmatrix} 1 & 0 \\ 0 & -1 \end{bmatrix}$

$$A'(-2, -1),\ B'(-1, -4),$$
$$C'(2, -2);\ \text{reflection across the}$$
x-axis

Solve.

9. a. Natalie drew a figure with vertices *H*(−3, −2), *O*(−3, 3), *U*(0, 5), *S*(3, 3), *E*(3, −2) to use as a pattern on a sweatshirt. Write a matrix that defines the figure.

$$\begin{bmatrix} -3 & -3 & 0 & 3 & 3 \\ -2 & 3 & 5 & 3 & -2 \end{bmatrix}$$

b. Natalie wants to enlarge the figure by a factor of 5. Describe a method she can use.

Multiply each entry in the matrix by 5.

c. What are the coordinates of Natalie's enlarged figure?

H' **(−15, −10)** O' **(−15, 15)** U' **(0, 25)** S' **(15, 15)** E' **(15, −10)**

Holt Algebra 2

Practice B
Determinants and Cramer's Rule

Find the determinant of each matrix.

1. $\begin{bmatrix} 8 & 2 \\ 4 & -1 \end{bmatrix}$

2. $\begin{bmatrix} -6 & 3 \\ 9 & -5 \end{bmatrix}$

3. $\begin{bmatrix} -2 & 8 \\ -3 & 7 \end{bmatrix}$

$\underline{\hspace{3em} -16 \hspace{3em}}$ $\underline{\hspace{3em} 3 \hspace{3em}}$ $\underline{\hspace{3em} 10 \hspace{3em}}$

4. $\begin{bmatrix} 1 & 0 & -1 \\ 5 & -2 & 0 \\ 1 & 6 & 2 \end{bmatrix}$

5. $\begin{bmatrix} 0 & -4 & 5 \\ 2 & 4 & 3 \\ 1 & 1 & -1 \end{bmatrix}$

6. $\begin{bmatrix} -4 & 3 & 1 \\ 7 & -2 & 0 \\ 1 & -1 & 2 \end{bmatrix}$

$\underline{\hspace{3em} -36 \hspace{3em}}$ $\underline{\hspace{3em} -30 \hspace{3em}}$ $\underline{\hspace{3em} -31 \hspace{3em}}$

Use Cramer's rule to solve each system of equations.

7. $\begin{cases} 2x + 3y = -1 \\ 3x + 2y = 16 \end{cases}$

8. $\begin{cases} 4x - 3y = 9 \\ 3x + 2y = 28 \end{cases}$

9. $\begin{cases} 8x - 3y = 20 \\ 3x - 2y = 11 \end{cases}$

$\underline{\hspace{2em} (10, -7) \hspace{2em}}$ $\underline{\hspace{2em} (6, 5) \hspace{2em}}$ $\underline{\hspace{2em} (1, -4) \hspace{2em}}$

10. $\begin{cases} 4y = -5x + 33 \\ 2y = 3x - 11 \end{cases}$

11. $\begin{cases} 27 + 4y = 3x \\ y = \frac{1}{3}x - 8 \end{cases}$

12. $\begin{cases} 7 - 5y + 4x = 0 \\ 16 - 2y - 5x = 0 \end{cases}$

$\underline{\hspace{2em} (5, 2) \hspace{2em}}$ $\underline{\hspace{2em} (-3, -9) \hspace{2em}}$ $\underline{\hspace{2em} (2, 3) \hspace{2em}}$

Solve.

13. On Monday, Marla babysat for 4 hours, did yard work for 2 hours, and earned a total of $41. On Friday, she babysat for 5 hours, did yard work for 3 hours, and earned a total of $55.

   a. Write a system of equations.
      Let x = Marla's hourly rate for babysitting,
      and y = her hourly rate for yard work.

$$\begin{cases} 4x + 2y = 41 \\ 5x + 3y = 55 \end{cases}$$

   b. Write the coefficient matrix. Evaluate
      its determinant.

$$\begin{bmatrix} 4 & 2 \\ 5 & 3 \end{bmatrix}; \det = \begin{vmatrix} 4 & 2 \\ 5 & 3 \end{vmatrix} = 2$$

   c. Use Cramer's rule to find x and y.

$$x = 6.5; \; y = 7.5$$

   d. What is Marla's hourly rate for each activity?

Babysitting: $6.50, yard work: $7.50

Holt Algebra 2

LESSON 4-5

Practice B
Matrix Inverses and Solving Systems

Determine whether the given matrices are inverses.

1. $\begin{bmatrix} -5 & 0 \\ 4 & 1 \end{bmatrix} \begin{bmatrix} -0.2 & 0 \\ 0.8 & 1 \end{bmatrix}$

   Yes

2. $\begin{bmatrix} 1 & -4 \\ -2 & 3 \end{bmatrix} \begin{bmatrix} -0.6 & -0.8 \\ -0.4 & -0.2 \end{bmatrix}$

   Yes

3. $\begin{bmatrix} 2 & -3 \\ -1 & 1 \end{bmatrix} \begin{bmatrix} -1 & -3 \\ -1 & -2 \end{bmatrix}$

   Yes

Find the inverse of the matrix, if it is defined.

4. $\begin{bmatrix} 1 & 0 \\ 4 & -1 \end{bmatrix}$

   $\begin{bmatrix} 1 & 0 \\ 4 & -1 \end{bmatrix}$

5. $\begin{bmatrix} 5 & 2 \\ 7 & 3 \end{bmatrix}$

   $\begin{bmatrix} 3 & -2 \\ -7 & 5 \end{bmatrix}$

6. $\begin{bmatrix} 8 & 4 \\ -5 & -3 \end{bmatrix}$

   $\begin{bmatrix} \dfrac{3}{4} & 1 \\ -\dfrac{5}{4} & -2 \end{bmatrix}$

7. $\begin{bmatrix} 3 & -3 \\ -2 & 1 \end{bmatrix}$

   $\begin{bmatrix} -\dfrac{1}{3} & -1 \\ -\dfrac{2}{3} & -1 \end{bmatrix}$

8. $\begin{bmatrix} -4 & 4 \\ 5 & -4 \end{bmatrix}$

   $\begin{bmatrix} 1 & 1 \\ \dfrac{5}{4} & 1 \end{bmatrix}$

9. $\begin{bmatrix} 6 & -6 \\ 1 & -1 \end{bmatrix}$

   The inverse does not exist.

Write the matrix equation for the system, and solve.

10. $\begin{cases} 3x + 2y = -5 \\ 4x + 3y = -9 \end{cases}$

    $\begin{bmatrix} 3 & 2 \\ 4 & 3 \end{bmatrix} \begin{bmatrix} x \\ y \end{bmatrix} = \begin{bmatrix} -5 \\ -9 \end{bmatrix}$; $(3, -7)$

11. $\begin{cases} -6x + 4y = 8 \\ 5x - 3y = -5 \end{cases}$

    $\begin{bmatrix} -6 & 4 \\ 5 & -3 \end{bmatrix} \begin{bmatrix} x \\ y \end{bmatrix} = \begin{bmatrix} 8 \\ -5 \end{bmatrix}$; $(2, 5)$

12. $\begin{cases} 4x + 5y = 0 \\ 5x + 3y = 13 \end{cases}$

    $\begin{bmatrix} 4 & 5 \\ 5 & 3 \end{bmatrix} \begin{bmatrix} x \\ y \end{bmatrix} = \begin{bmatrix} 0 \\ 13 \end{bmatrix}$; $(5, -4)$

13. $\begin{cases} 5x - 3y = 8 \\ 6x - 5y = 4 \end{cases}$

    $\begin{bmatrix} 5 & -3 \\ 6 & -5 \end{bmatrix} \begin{bmatrix} x \\ y \end{bmatrix} = \begin{bmatrix} 8 \\ 4 \end{bmatrix}$; $(4, 4)$

Solve.

14. Keith paid $39 for 3 pounds of pistachios and 2 pounds of cashews.
    Tracey paid $23 for 2 pounds of pistachios and 1 pound of cashews.

    a. Write a system of equations. Let x = the cost
    of a pound of pistachios, and y = the cost of a
    pound of cashews.

    $\begin{cases} 3x + 2y = 39 \\ 2x + y = 23 \end{cases}$

    $\begin{bmatrix} 3 & 2 \\ 2 & 1 \end{bmatrix} \begin{bmatrix} x \\ y \end{bmatrix} = \begin{bmatrix} 29 \\ 23 \end{bmatrix}$;

    **pistachios: $7 per pound,
    cashews: $9 per pound**

    b. Write the matrix equation and solve.

Holt Algebra 2

Practice B
Row Operations and Augmented Matrices

Write the augmented matrix for each system of equations.

1. $\begin{cases} 2x + 1 = y \\ x + y + z = 1 \\ 4y + 5z = 3 \end{cases}$

$\begin{bmatrix} 2 & -1 & 0 & -1 \\ 1 & 1 & 1 & 1 \\ 0 & 4 & 5 & 3 \end{bmatrix}$

2. $\begin{cases} 3x = 2y + 4 \\ x - y = 3z \\ 2y + 8z = x \end{cases}$

$\begin{bmatrix} 3 & -2 & 0 & 4 \\ 1 & -1 & -3 & 0 \\ -1 & 2 & 8 & 0 \end{bmatrix}$

3. $\begin{cases} x + z = 1 \\ 3x - 5y = 12 \\ 2y - 3z = 9 \end{cases}$

$\begin{bmatrix} 1 & 0 & 1 & 1 \\ 3 & -5 & 0 & 12 \\ 0 & 2 & -3 & 9 \end{bmatrix}$

Write the augmented matrix, and use row reduction to solve.

4. $\begin{cases} 4x + 3y = -11 \\ 2x - 3y = 17 \end{cases}$

$\begin{bmatrix} 4 & 3 & -11 \\ 2 & -3 & 17 \end{bmatrix}$; $(1, -5)$

5. $\begin{cases} 3x + 7y = -1 \\ 6x + 11y = 10 \end{cases}$

$\begin{bmatrix} 3 & 7 & -1 \\ 6 & 11 & 10 \end{bmatrix}$; $(9, -4)$

6. $\begin{cases} 2x = 3y - 1 \\ 5x - 12y = 2 \end{cases}$

$\begin{bmatrix} 2 & -3 & -1 \\ 5 & -12 & 2 \end{bmatrix}$; $(-2, -1)$

7. $\begin{cases} x + 6y = 0 \\ 2x + 9y = -3 \end{cases}$

$\begin{bmatrix} 1 & 6 & 0 \\ 2 & 9 & -3 \end{bmatrix}$; $(-6, 1)$

Solve.

8. Dimitri has $4.95 in dimes and quarters. He has 3 fewer dimes than quarters.

a. Write a system of equations.
Let d = the number of dimes and
q = the number of quarters.

$\begin{cases} 10d + 25q = 495 \\ d = q - 3 \end{cases}$

b. Write the augmented matrix for the system.

$\begin{bmatrix} 10 & 25 & 495 \\ 1 & -1 & -3 \end{bmatrix}$

c. How many of each coin does Dimitri have?

12 dimes and 15 quarters

9. Clara has a bag of 60 coins with a value of $2.00. The coins are all pennies and nickels. How many of each coin are in the bag?

35 nickels and 25 pennies

Holt Algebra 2

LESSON 5-1 Practice B

Using Transformations to Graph Quadratic Functions

Graph the function by using a table.

1. $f(x) = x^2 + 2x - 1$

x	$f(x) = x^2 + 2x - 1$	$(x, f(x))$
-2	-1	$(-2, -1)$
-1	-2	$(-1, -2)$
0	-1	$(0, -1)$
1	2	$(1, 2)$
2	7	$(2, 7)$

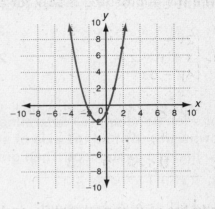

Using the graph of $f(x) = x^2$ as a guide, describe the transformations, and then graph each function. Label each function on the graph.

2. $h(x) = (x - 2)^2 + 2$

Translated 2 units right, 2 units up

3. $h(x) = -(3x)^2$

Reflected across the x-axis and horizontal compression by a factor of 3

4. $h(x) = \left(\frac{1}{2}x\right)^2$

Horizontal stretch by a factor of 2

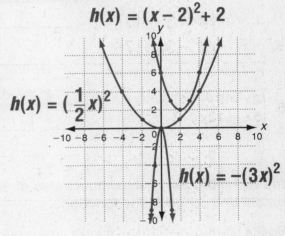

Use the description to write a quadratic function in vertex form.

5. The parent function $f(x) = x^2$ is reflected across the x-axis, horizontally stretched by a factor of 3 and translated 2 units down to create function g.

$$g(x) = -\left(\frac{1}{3}x\right)^2 - 2$$

6. A ball dropped from the top of tower A can be modeled by the function $h(t) = -9.8t^2 + 400$, where t is the time after it is dropped and $h(t)$ is its height at that time. A ball dropped from the top of tower B can be modeled by the function $h(t) = -9.8t^2 + 200$. What transformation describes this change? What does this transformation mean?

Vertical translation; possible answer: at a given time a ball dropped from tower A will be 200 feet higher than a ball dropped from tower B at the same time. Tower A is 200 feet taller than tower B.

Holt Algebra 2

Name _____ Date _____ Class _____

Identify the axis of symmetry for the graph of each function.

1. $g(x) = x^2 - 4x + 2$ **2.** $h(x) = -8x^2 + 12x - 11$ **3.** $k(x) = -4(x + 3)^2 + 9$

      $x = 2$ **$x = \dfrac{3}{4}$**   **$x = -3$**

For each function, (a) determine whether the graph opens upward or downward, (b) find the axis of symmetry, (c) find the vertex, and (d) find the y-intercept. Then graph the function.

4. $f(x) = -x^2 + 3x + 1$

   a. Upward or downward   **Downward**

   b. Axis of symmetry   **$x = 1.5$**

   c. Vertex   **(1.5, 3.25)**

   d. y-intercept   **1**

5. $g(x) = 2x^2 + 4x - 2$

   a. Upward or downward   **Upward**

   b. Axis of symmetry   **$x = -1$**

   c. Vertex   **(−1, −4)**

   d. y-intercept   **−2**

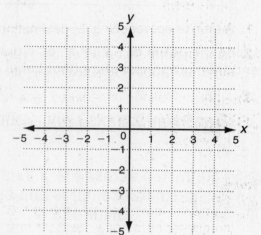

Find the minimum or maximum value of each function. Then state the domain and range of the function.

6. $g(x) = x^2 - 2x + 1$ **7.** $h(x) = -5x^2 + 15x - 3$

   Minimum: 0; domain: all real   **Maximum: 8.25; domain: all real**
   numbers; range: $\{y \mid y \geq 0\}$ **numbers; range: $\{y \mid y \leq 8.25\}$**

Solve.

8. A record label uses the following function to model the sales of a new release.

$$a(t) = -90t^2 + 8100t$$

The number of albums sold is a function of time, t, in days. On which day were the most albums sold? What is the maximum number of albums sold on that day?

          Day 45; 182,250 records

                Holt Algebra 2

Name _____ Date _____ Class _____

Practice B
Solving Quadratic Equations by Graphing and Factoring

Find the zeros of each function by using a graph and a table.

1. $f(x) = x^2 + 5x + 6$

x	−4	−3	−2	−1	0
f(x)	2	0	0	2	6

<u>　　　−2 and −3　　　</u>

2. $g(x) = -x^2 + 4x + 5$

x	−2	0	2	4	6
f(x)	−7	5	9	5	−7

<u>　　−1 and 5　　</u>

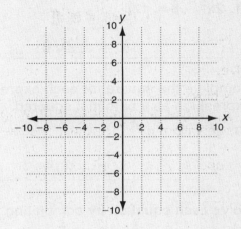

Find the zeros of each function by factoring.

3. $h(x) = -x^2 - 6x - 9$ **4.** $f(x) = 2x^2 + 9x + 4$ **5.** $g(x) = x^2 + x - 20$

<u>　　　−3　　　</u>   <u>　　−0.5, −4　　</u>   <u>　　−5, 4　　</u>

Find the roots of each equation by factoring.

6. $12x = 9x^2 + 4$ **7.** $16x^2 = 9$

<u>　　$\dfrac{2}{3}$　　</u>   <u>　　−0.75, 0.75　　</u>

Write a quadratic function in standard form for each given set of zeros.

8. −2 and 7   **9.** 1 and −8

<u>　$f(x) = x^2 - 5x - 14$　</u>   <u>　$f(x) = x^2 + 7x - 8$　</u>

Solve.

10. The quadratic function that approximates the height of a javelin throw is $h(t) = -0.08t^2 + 4.48$, where t is the time in seconds after it is thrown and h is the javelin's height in feet. How long will it take for the javelin to hit the ground?

<u>　　About 7.5 s　　</u>

Holt Algebra 2

LESSON 5-4

Practice B
Completing the Square

Solve each equation.

1. $2x^2 - 6 = 42$

$$x = \pm 2\sqrt{6}$$

2. $x^2 - 14x + 49 = 18$

$$x = 7 \pm 3\sqrt{2}$$

Complete the square for each expression. Write the resulting expression as a binomial squared.

3. $x^2 - 4x + \underline{\;4\;}$

$$(x - 2)^2$$

4. $x^2 + 12x + \underline{\;36\;}$

$$(x + 6)^2$$

Solve each equation by completing the square.

5. $2d^2 = 8 + 10d$

$$d = \frac{5}{2} \pm \frac{\sqrt{41}}{2}$$

6. $x^2 + 2x = 3$

$$x = -3, 1$$

7. $-3x^2 + 18x = -30$

$$x = 3 \pm \sqrt{19}$$

8. $4x^2 = -12x + 4$

$$x = -\frac{3}{2} \pm \frac{\sqrt{13}}{2}$$

Write each function in vertex form, and identify its vertex.

9. $f(x) = x^2 - 6x - 2$

$$f(x) = (x - 3)^2 - 11; \ (3, -11)$$

10. $f(x) = x^2 - 4x + 1$

$$g(x) = (x - 2)^2 - 3; \ (2, -3)$$

11. $h(x) = 3x^2 - 6x - 15$

$$h(x) = 3(x - 1)^2 - 18; \ (1, -18)$$

12. $f(x) = -2x^2 - 16x + 4$

$$f(x) = -2(x + 4)^2 + 36; \ (-4, 36)$$

Solve.

13. Nathan made a triangular pennant for the band booster club. The area of the pennant is 80 square feet. The base of the pennant is 12 feet shorter than the height.

a. What are the lengths of the base and height of the pennant?

$$\text{Base} = 8 \text{ ft, height} = 20 \text{ ft}$$

b. What are the dimensions of the pennant if the base is only 6 feet shorter than the height?

$$\text{Base} = 10 \text{ ft, height} = 16 \text{ ft}$$

Holt Algebra 2

Practice B

Complex Numbers and Roots

Express each number in terms of *i*.

1. $\sqrt{-32}$

2. $2\sqrt{-18}$

3. $\sqrt{-\dfrac{1}{9}}$

$4i\sqrt{2}$

$6i\sqrt{2}$

$\dfrac{1}{3}i$

Solve each equation.

4. $3x^2 + 81 = 0$

5. $4x^2 = -28$

$x = \pm 3i\sqrt{3}$

$x = \pm i\sqrt{7}$

6. $\dfrac{1}{4}x^2 + 12 = 0$

7. $6x^2 = -126$

$x = \pm 4i\sqrt{3}$

$x = \pm i\sqrt{21}$

Find the values of *x* and *y* that make each equation true.

8. $2x - 20i = 8 - (4y)i$

9. $5i - 6x = (10y)i + 2$

$x = 4,\ y = 5$

$x = -\dfrac{1}{3},\ y = \dfrac{1}{2}$

Find the zeros of each function.

10. $f(x) = x^2 - 2x + 4$

11. $g(x) = x^2 + 6x + 14$

$x = 1 \pm i\sqrt{3}$

$x = -3 \pm i\sqrt{5}$

Find each complex conjugate.

12. $i - 3$

13. $3i - 4$

14. $11i$

$-3 - i$

$-4 - 3i$

$-11i$

Solve.

15. The impedance of an electrical circuit is a way of measuring how much the circuit impedes the flow of electricity. The impedance can be a complex number. A circuit is being designed that must have an impedance that satisfies the function $f(x) = 2x^2 - 12x + 40$, where *x* is a measure of the impedance. Find the zeros of the function.

$3 \pm i\sqrt{11}$

Holt Algebra 2

LESSON
5-6

Practice B
The Quadratic Formula

Find the zeros of each function by using the Quadratic Formula.

1. $f(x) = x^2 + 10x + 9$

$$x = -9, -1$$

2. $g(x) = 2x^2 + 4x - 12$

$$x = -1 \pm \sqrt{7}$$

3. $h(x) = 3x^2 - 3x + \dfrac{3}{4}$

$$x = 0.5$$

4. $f(x) = x^2 + 2x - 3$

$$x = -3, 1$$

5. $g(x) = 2x^2 + 3x + 1$

$$x = -1, -0.5$$

6. $g(x) = x^2 + 5x + -3$

$$x = \dfrac{-5 \pm \sqrt{37}}{2}$$

Find the type and number of solutions for each equation.

7. $x^2 - 3x = -8$

Two nonreal solutions

8. $x^2 + 4x = -3$

Two real solutions

9. $2x^2 - 12x = -18$

One real solution

Solve.

10. A newspaper delivery person in a car is tossing folded newspapers from the car window to driveways. The speed of the car is 30 feet per second, and the driver does not slow down. The newspapers are tossed horizontally from a height of 4 feet above the ground. The height of the papers as they are thrown can be modeled by $y = -16t^2 + 4$, and the distance they travel to the driveway is $d = 30t$.

a. How long does it take for a newspaper to land?

0.5 s

b. From how many feet before the driveway must the papers be thrown?

15 ft

c. The delivery person starts to throw the newspapers at an angle and the height of the papers as they travel can now be modeled by $y = -16t^2 + 12t + 4$. How long does it take the papers to reach the ground now?

1 s

Holt Algebra 2

Name _____ Date _____ Class _____

Practice B
Solving Quadratic Inequalities

Graph each inequality.

1. $y < x^2 - 2x + 6$

2. $y > 2x^2 - x - 7$

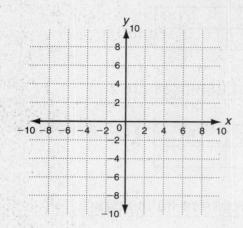

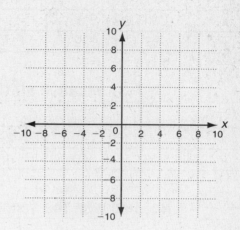

Solve each inequality by using tables or graphs.

3. $x^2 + 3x - 14 \leq 14$

$$-7 \leq x \leq 4$$

4. $x^2 - 9x > -18$

$$x < 3 \text{ or } x > 6$$

Solve each inequality by using algebra.

5. $x^2 - x - 3 > x$

$$x < -1 \text{ or } x > 3$$

6. $x^2 + 6x + 3 < -2$

$$-5 < x < -1$$

7. $3 \leq x^2 - 8x + 15$

$$x < 2 \text{ or } x > 6$$

8. $3x^2 + x + 8 \leq 12$

$$-\frac{4}{3} < x < 1$$

Solve.

9. An online music service that sells song downloads models its profit using the function $P(d) = -5d^2 + 450d - 1000$, where d is the number of downloads sold and P is the profit. How many downloads does it need to sell to make a profit of more than \$8000?

More than 30 but fewer than 60

Holt Algebra 2

Practice B

LESSON 5-8 *Curve Fitting with Quadratic Models*

Determine whether each data set could represent a quadratic function. Explain.

1.

x	−1	0	1	2	3
y	35	22	11	2	−5

Yes, because all the second differences are 2

2.

x	−2	0	2	4	6
y	18	10	6	2	1

No, because the second differences are not constant

Write a quadratic equation that fits each set of points.

3. $(0, -8)$, $(2, 0)$, and $(-3, -5)$

$$f(x) = x^2 + 2x - 8$$

4. $(-1, -16)$, $(2, 5)$, and $(5, 8)$

$$f(x) = -x^2 + 8x - 7$$

5. $(-2, 6)$, $(0, -6)$, and $(3, -9)$

$$f(x) = x^2 - 4x - 6$$

6. $(1, 4)$, $(-2, 13)$, and $(0, 3)$

$$f(x) = 2x^2 - x + 3$$

Solve.

7. The data table shows the energy, *E,* of a certain object in joules at a given velocity, *v,* in meters per second.

Energy (joules)	4.5	12.5	24.5	40.5
Velocity (m/s)	1.5	2.5	3.5	4.5

a. Find the quadratic relationship between the energy and velocity of the object.

   $E = 2v^2$

b. What is the energy of an object with a speed of 5 m/s?

   50 joules

c. What is the velocity of the object if the energy is 128 joules?

8 m/s

Holt Algebra 2

Name _____ Date _____ Class _____

Practice B

5-9 *Operations with Complex Numbers*

Graph each complex number.

Imaginary axis

1. -6

2. $4i$

3. $6 + 7i$

4. $-8 - 5i$

5. $-3i$

Find each absolute value.

6. $|4 + 2i|$

$$2\sqrt{5}$$

7. $|5 - i|$

$$\sqrt{26}$$

8. $|-3i|$

$$3$$

Add or subtract. Write the result in the form $a + bi$.

9. $(-1 + 2i) + (6 - 9i)$

$$5 - 7i$$

10. $(3 - 3i) - (4 + 7i)$

$$-1 - 10i$$

11. $(-5 + 2i) + (-2 + 8i)$

$$-7 + 10i$$

Multiply. Write the result in the form $a + bi$.

12. $3i(2 - 3i)$

$$9 + 6i$$

13. $(4 + 5i)(2 + i)$

$$3 + 14i$$

14. $(-1 + 6i)(3 - 2i)$

$$9 + 20i$$

Simplify.

15. $\dfrac{2 + 4i}{3i}$

$$\dfrac{4}{3} - \dfrac{2}{3}i$$

16. $\dfrac{3 + 2i}{4 + i}$

$$\dfrac{14}{17} + \dfrac{5}{17}i$$

17. $2i^{11}$

$$-2i$$

Solve.

18. In electronics, the total resistance to the flow of electricity in a circuit is called the impedance, Z. Impedance is represented by a complex number. The total impedance in a series circuit is the sum of individual impedances. The impedance in one part of a circuit is $Z_1 = 3 + 4i$. In another part of a circuit, the impedance is $Z_1 = 5 - 2i$. What is the total impedance of the circuit?

$$8 + 2i$$

Holt Algebra 2

Name _____ Date _____ Class _____

Practice B
Polynomials

Identify the degree of each monomial.

1. $6x^2$

2. $3p^3m^4$

3. $2x^8y^3$

_____2_____   _____7_____   _____11_____

Rewrite each polynomial in standard form. Then identify the leading coefficient, degree, and number of terms. Name the polynomial.

4. $6 + 7x - 4x^3 + x^2$

$-4x^3 + x^2 + 7x + 6$; -4; 3; 4; cubic polynomial with 4 terms

5. $x^2 - 3 + 2x^5 + 7x^4 - 12x$

$2x^5 + 7x^4 + x^2 - 12x - 3$; 2; 5; 5; quintic polynomial with 5 terms

Add or subtract. Write your answer in standard form.

6. $(2x^2 - 2x + 6) + (11x^3 - x^2 - 2 + 5x)$

$11x^3 + x^2 + 3x + 4$

7. $(x^2 - 8) - (3x^3 + 6x - 4 + 9x^2)$

$-3x^3 - 8x^2 - 6x - 4$

8. $(5x^4 + x^2) + (7 + 9x^2 - 2x^4 + x^3)$

$3x^4 + x^3 + 10x^2 + 7$

9. $(12x^2 + x) - (6 - 9x^2 + x^7 - 8x)$

$-x^7 + 21x^2 + 9x - 6$

Graph each polynomial function on a calculator. Describe the graph, and identify the number of real zeros.

10. $f(x) = x^3 + 2x^2 - 3$

From left to right, the graph increases, decreases slightly, and then increases again. It crosses the x-axis once, so there is 1 real zero.

11. $f(x) = x^4 - 5x^2 + 1$

From left to right, the graph alternately decreases and increases, changing direction 3 times. It crosses the x-axis 4 times, so there are 4 real zeros.

Solve.

12. The height, h, in feet, of a baseball after being struck by a bat can be approximated by $h(t) = -16t^2 + 100t + 5$, where t is measured in seconds.

  a. Evaluate $h(t)$ for $t = 3$ and $t = 5$.

  161 ft and 105 ft

  b. Describe what the values of the function from part a represent.

  The height of the baseball 3 s after being hit by the bat and the height of the baseball 5 s after being hit by the bat

Holt Algebra 2

Practice B

LESSON 6-2

Multiplying Polynomials

Find each product.

1. $4x^2(3x^2 + 1)$

$$12x^4 + 4x^2$$

2. $-9x(x^2 + 2x + 4)$

$$-9x^3 - 18x^2 - 36x$$

3. $-6x^2(x^3 + 7x^2 - 4x + 3)$

$$-6x^5 - 42x^4 + 24x^3 - 18x^2$$

4. $x^3(-4x^3 + 10x^2 - 7x + 2)$

$$-4x^6 + 10x^5 - 7x^4 + 2x^3$$

5. $-5m^3(7n^4 - 2mn^3 + 6)$

$$-35m^3n^4 + 10m^4n^3 - 30m^3$$

6. $(x + 2)(y^2 + 2y - 12)$

$$xy^2 + 2xy - 12x + 2y^2 + 4y - 24$$

7. $(p + q)(4p^2 - p - 8q^2 - q)$

$$4p^3 - p^2 + 4p^2q - 2pq - 8pq^2 - q^2 - 8q^3$$

8. $(2x^2 + xy - y)(y^2 + 3x)$

$$2x^2y^2 + 6x^3 + xy^3 + 3x^2y - y^3 - 3xy$$

Expand each expression.

9. $(3x - 1)^3$

$$27x^3 - 27x^2 + 9x - 1$$

10. $(x - 4)^4$

$$x^4 - 16x^3 + 96x^2 - 256x + 256$$

11. $3(a - 4b)^2$

$$3a^2 - 24ab + 48b^2$$

12. $5(x^2 - 2y^3)^3$

$$5x^6 - 30x^4y + 60x^2y^2 - 40y^3$$

Solve.

13. A biologist has found that the number of branches on a certain rare tree in its first few years of life can be modeled by the polynomial $b(y) = 4y^2 + y$. The number of leaves on each branch can be modeled by the polynomial $l(y) = 2y^3 + 3y^2 + y$, where y is the number of years after the tree reaches a height of 6 feet. Write a polynomial describing the total number of leaves on the tree.

$$8y^5 + 14y^4 + 7y^3 + y^2$$

Holt Algebra 2

Name _____ Date _____ Class _____

Practice B
Dividing Polynomials

Divide by using long division.

1. $(x^2 - x - 6) \div (x - 3)$

$$x + 2$$

2. $(2x^3 - 10x^2 + x - 5) \div (x - 5)$

$$2x^2 + 1$$

3. $(-3x^2 + 20x - 12) \div (x - 6)$

$$-3x + 2$$

4. $(3x^3 + 9x^2 - 14) \div (x + 3)$

$$3x^2 - \frac{14}{x + 3}$$

Divide by using synthetic division.

5. $(3x^2 - 8x + 4) \div (x - 2)$

$$3x - 2$$

6. $(5x^2 - 4x + 12) \div (x + 3)$

$$5x - 19 + \frac{69}{x + 3}$$

7. $(9x^2 - 7x + 3) \div (x - 1)$

$$9x + 2 + \frac{5}{x - 1}$$

8. $(-6x^2 + 5x - 10) \div (x + 7)$

$$-6x + 47 - \frac{339}{x + 7}$$

Use synthetic substitution to evaluate the polynomial for the given value.

9. $P(x) = 4x^2 - 9x + 2$ for $x = 3$

$$P(3) = 11$$

10. $P(x) = -3x^2 + 10x - 4$ for $x = -2$

$$P(-2) = -36$$

Solve.

11. The total number of dollars donated each year to a small charitable organization has followed the trend $d(t) = 2t^3 + 10t^2 + 2000t + 10{,}000$, where d is dollars and t is the number of years since 1990. The total number of donors each year has followed the trend $p(t) = t^2 + 1000$. Write an expression describing the average number of dollars per donor.

$$2t + 10$$

Holt Algebra 2

LESSON 6-4 Practice B
Factoring Polynomials

Determine whether the given binomial is a factor of the polynomial
$P(x)$.

1. $(x - 4)$; $P(x) = x^2 + 8x - 48$

Yes

2. $(x + 5)$; $P(x) = 2x^2 - 6x - 1$

No

3. $(x - 6)$; $P(x) = -2x^2 + 15x - 18$

Yes

4. $(x + 3)$; $P(x) = 2x^2 - x + 7$

No

Factor each expression.

5. $2x^4 + 2x^3 - x^2 - x$

$x(2x - 1)(x + 1)$

6. $4x^3 + x^2 - 8x - 2$

$(4x + 1)(x^2 - 2)$

7. $5x^6 - 5x^4 + x^3 - x$

$x(5x^3 + 1)(x^2 - 1)$

8. $2x^4 + 54x$

$2x(x + 3)(x^2 - 3x + 9)$

9. $64x^3 - 1$

$(4x - 1)(16x^2 + 4x + 1)$

10. $3x^4 + 24x$

$3x(x + 2)(x^2 - 2x + 4)$

Solve.

11. Since 2006, the water level in a certain pond
has been modeled by the polynomial
$d(x) = -x^3 + 16x^2 - 74x + 140$, where the depth
d, is measured in feet over x years. Identify the
year that the pond will dry up. Use the graph to
factor $d(x)$.

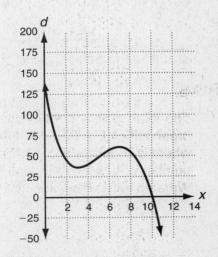

$2016; -(x - 10)(x^2 - 6x + 14)$

43

Holt Algebra 2

LESSON 6-5 Practice B
Finding Real Roots of Polynomial Equations

Solve each polynomial equation by factoring.

1. $9x^3 - 3x^2 - 3x + 1 = 0$

$$\frac{1}{3}, \frac{\sqrt{3}}{3}, -\frac{\sqrt{3}}{3}$$

2. $x^5 - 2x^4 - 24x^3 = 0$

$$-4, 0, 6$$

3. $3x^5 + 18x^4 - 21x^3 = 0$

$$-7, 0, 1$$

4. $-x^4 + 2x^3 + 8x^2 = 0$

$$-2, 0, 4$$

Identify the roots of each equation. State the multiplicity of each root.

5. $x^3 + 3x^2 + 3x + 1 = 0$

$$x = -1 \text{ with multiplicity } 3$$

6. $x^3 + 5x^2 - 8x - 48 = 0$

$x = 3$ with multiplicity 1; $x = -4$ with multiplicity 2

Identify all the real roots of each equation.

7. $x^3 + 10x^2 + 17x = 28$

$$-4, 1, -7$$

8. $3x^3 + 10x^2 - 27x = 10$

$$-5, -\frac{1}{3}, 2$$

Solve.

9. An engineer is designing a storage compartment in a spacecraft. The compartment must be 2 meters longer than it is wide and its depth must be 1 meter less than its width. The volume of the compartment must be 8 cubic meters.

a. Write an equation to model the volume of the compartment.

$$x^3 + x^2 - 2x - 8 = 0$$

b. List all possible rational roots. $\pm1, \pm2, \pm4, \pm8$

c. Use synthetic division to find the roots of the polynomial equation. Are the roots all rational numbers?

$2, \dfrac{-3 \pm i\sqrt{7}}{2}$; **no, 2 of the roots are irrational numbers.**

d. What are the dimensions of the storage compartment?   **2 m wide, 4 m long, and 1 m deep**

Holt Algebra 2

LESSON 6-6 **Practice B**
Fundamental Theorem of Algebra

Write the simplest polynomial function with the given roots.

1. 1, 4, and -3

$$P(x) = x^3 - 2x^2 - 11x + 12$$

2. $\frac{1}{2}$, 5, and -2

$$P(x) = x^3 - \frac{7}{2}x^2 - \frac{17}{2}x + 5$$

3. $2i$, $\sqrt{3}$, and 4

$$P(x) = x^5 - 4x^4 + x^3 - 4x^2 \\ - 12x + 48$$

4. $\sqrt{2}$, -5, and $-3i$

$$P(x) = x^5 + 5x^4 + 7x^3 + 35x^2 \\ - 18x - 90$$

Solve each equation by finding all roots.

5. $x^4 - 2x^3 - 14x^2 - 2x - 15 = 0$

$$x = i, -i, -3, \text{ and } 5$$

6. $x^4 - 16 = 0$

$$x = 2, -2, 2i, \text{ and } -2i$$

7. $x^4 + 4x^3 + 4x^2 + 64x - 192 = 0$

$$x = -4i, 4i, 2, \text{ and } -6$$

8. $x^3 + 3x^2 + 9x + 27 = 0$

$$x = -3i, 3i, \text{ and } -3$$

Solve.

9. An electrical circuit is designed such that its output voltage, V, measured in volts, can be either positive or negative. The voltage of the circuit passes through zero at $t = 1$, 2, and 7 seconds. Write the simplest polynomial describing the voltage $V(t)$.

$$V(t) = t^3 - 10t^2 + 23t - 14$$

Holt Algebra 2

LESSON 6-7 Practice B
Investigating Graphs of Polynomial Functions

Identify the leading coefficient, degree, and end behavior.

1. $P(x) = 2x^5 - 6x^3 + x^2 - 2$

   2; 5; as $x \to +\infty$, $P(x) \to +\infty$; and as $x \to -\infty$, $P(x) \to -\infty$

2. $Q(x) = -4x^2 + x - 1$

   -4; 2; as $x \to -\infty$, $Q(x) \to -\infty$; and as $x \to +\infty$, $Q(x) \to -\infty$

Identify whether the function graphed has an odd or even degree and a positive or negative leading coefficient.

3.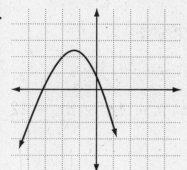

Even; negative

4.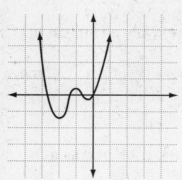

Even; positive

5.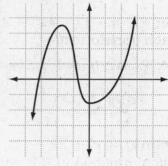

Odd; positive

Graph the function $P(x) = x^3 + 6x^2 + 5x - 12$.

6. Identify the possible rational roots.

   $\pm 1, \pm 2, \pm 3, \pm 4, \pm 6, \pm 12$

7. Identify the zeros.

   $-4, -3$, and 1

8. Describe the end behavior of the function.

   As $x \to +\infty$, $P(x) \to +\infty$, and as $x \to -\infty$, $P(x) \to -\infty$

9. Sketch the graph of the function.

Solve.

10. The number, $N(y)$, of subscribers to a local magazine can be modeled by the function $N(y) = 0.1y^4 - 3y^3 + 10y^2 - 30y + 10,000$, where y is the number of years since the magazine was founded. Graph the polynomial on a graphing calculator and find the minimum number of subscribers and the year in which this occurs.

About 5400 in year 20

Holt Algebra 2

Practice B
Transforming Polynomial Functions

For $f(x) = x^3 + 1$, write the rule for each function and sketch its graph.

1. $g(x) = f(x + 4)$

   $g(x) = (x + 4)^3 + 1$

2. $g(x) = 3f(x)$

   $g(x) = 3x^3 + 3$

3. $g(x) = f\left(\frac{1}{2}x\right)$

   $g(x) = \left(\frac{1}{2}x\right)^3 + 1$

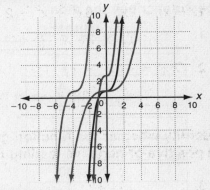

Let $f(x) = -x^3 + 4x^2 - 5x + 12$. Write a function $g(x)$ that performs each transformation.

4. Reflect $f(x)$ across the y-axis

   $g(x) = x^3 + 4x^2 + 5x + 12$

5. Reflect $f(x)$ across the x-axis

   $g(x) = x^3 - 4x^2 + 5x - 12$

Let $f(x) = x^3 + 2x^2 - 3x - 6$. Describe $g(x)$ as a transformation of $f(x)$ and graph.

6. $g(x) = \frac{1}{4}f(x)$

   Vertically compressed by a factor of 4

7. $g(x) = f(x - 6)$

   Translated 6 units right

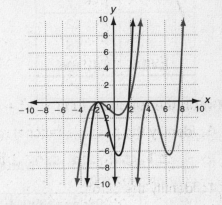

Write a function that transforms $f(x) = x^3 + 4x^2 - x + 5$ in each of the following ways. Support your solution by using a graphing calculator.

8. Move 6 units up and reflect across the y-axis.

   $-x^3 + 4x^2 + x + 11$

9. Compress vertically by a factor of 0.25 and move 3 units right.

   $\frac{1}{4}(x - 3)^3 + (x - 3)^2 - \frac{1}{4}(x - 3) + \frac{5}{4}$

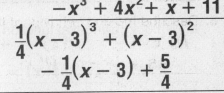

Solve.

10. The number of participants, N, in a new Internet political forum during each month of the first year can be modeled by $N(t) = 4t^2 - t + 2000$, where t is the number of months since January. In the second year, the number of forum participants doubled compared to the same month in the previous year. Write a function that describes the number of forum participants in the second year.

   $N(t) = 8t^2 - 2t + 4000$

Holt Algebra 2

LESSON 6-9 Practice B
Curve Fitting with Polynomial Models

Use finite differences to determine the degree of the polynomial that best describes the data.

1. _____ Quartic _____

x	y
0	4
1	14
2	24
3	30
4	30
5	24

2. _____ Cubic _____

x	y
−2	70
−1	35
0	15
1	7
2	8
3	15

3. _____ Quadratic _____

x	y
2	1
1	7
0	12
−1	16
−2	19
−3	21

4. _____ Cubic _____

x	y
−6	−31
−5	0
−4	16
−3	19
−2	11
−1	−6

Solve.

5. The data set shows the average price for a luxury commodity for the years since 1998.

Year	1998	1999	2000	2001	2002	2003	2004	2005
Price ($)	1000	2027	4472	7507	10,472	12,875	14,392	14,867

a. Write a polynomial function for the data.

$$f(x) = 7y^4 - 180y^3 + 1200y^2 + 1000$$

b. Predict the price of the item in 2008.

$11,000

Holt Algebra 2

LESSON 7-1

Practice B
Exponential Functions, Growth, and Decay

Tell whether the function shows growth or decay. Then graph.

1. $g(x) = -(2)^x$

2. $h(x) = -0.5(0.2)^x$

Growth

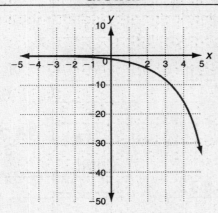

Decay

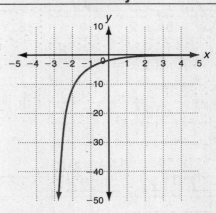

3. $j(x) = -2(0.5)^x$

4. $p(x) = 4(1.4)^x$

Decay

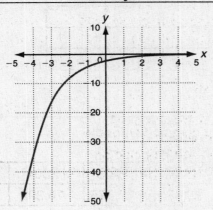

Growth

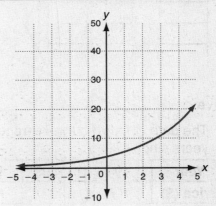

Solve.

5. A certain car depreciates about 15% each year.

   a. Write a function to model the depreciation in value for a car valued at $20,000.

   $$y = 20,000(0.85)^x$$

   b. Graph the function.

   c. Suppose the car was worth $20,000 in 2005. What is the first year that the value of this car will be worth less than half of that value?

   2010

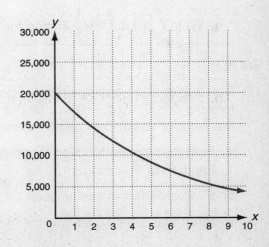

Holt Algebra 2

Name _____ Date _____ Class _____

Practice B
Inverses of Relations and Functions

Use inverse operations to write the inverse of each function.

1. $f(x) = 15x - 10$

$$f^{-1}(x) = \frac{x + 10}{15}$$

2. $f(x) = 10 - 4x$

$$f^{-1}(x) = -\frac{x - 10}{4}$$

3. $f(x) = 12 - 9x$

$$f^{-1}(x) = -\frac{x - 12}{9}$$

4. $f(x) = 5x + 2$

$$f^{-1}(x) = \frac{x - 2}{5}$$

5. $f(x) = x + 6$

$$f^{-1}(x) = x - 6$$

6. $f(x) = x + \frac{1}{2}$

$$f^{-1}(x) = x - \frac{1}{2}$$

7. $f(x) = -\frac{x}{12}$

$$f^{-1}(x) = -12x$$

8. $f(x) = \frac{x - 12}{4}$

$$f^{-1}(x) = 4x + 12$$

9. $f(x) = \frac{3x + 1}{6}$

$$f^{-1}(x) = \frac{6x - 1}{3}, \text{ or}$$

$$f^{-1}(x) = 2x - \frac{1}{3}$$

Graph each function. Then write and graph its inverse.

10. $f(x) = 2x - 4$

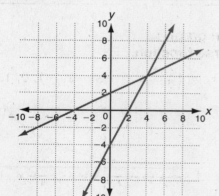

$$f^{-1}(x) = \frac{1}{2}x + 2$$

11. $f(x) = \frac{5}{2}x - 2$

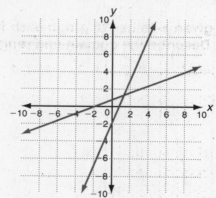

$$f^{-1}(x) = \frac{2}{5}(x + 2)$$

Solve.

12. Dan works at a hardware store. The employee discount is determined by the formula $d = 0.15(c - 10)$. Use the inverse of this function to find the cost of the item for which Dan received an $18.00 discount.

a. Find the inverse function that models cost as a function of the discount.

$$c = \frac{d + 1.5}{0.15}$$

b. Evaluate the inverse function for $d = 18$.

$$c = 130$$

c. What was Dan's final cost for this item?

$112

Holt Algebra 2

Practice B

LESSON 7-3

Logarithmic Functions

Write each exponential equation in logarithmic form.

1. $3^7 = 2187$

$\underline{\log_3 2187 = 7}$

2. $12^2 = 144$

$\underline{\log_{12} 144 = 2}$

3. $5^3 = 125$

$\underline{\log_5 125 = 3}$

Write each logarithmic equation in exponential form.

4. $\log_{10} 100{,}000 = 5$

$\underline{10^5 = 100{,}000}$

5. $\log_4 1024 = 5$

$\underline{4^5 = 1024}$

6. $\log_9 729 = 3$

$\underline{9^3 = 729}$

Evaluate by using mental math.

7. $\log 1{,}000{,}000$

$\underline{\qquad 6 \qquad}$

8. $\log 10$

$\underline{\qquad 1 \qquad}$

9. $\log 1$

$\underline{\qquad 0 \qquad}$

10. $\log_4 16$

$\underline{\qquad 2 \qquad}$

11. $\log_8 1$

$\underline{\qquad 0 \qquad}$

12. $\log_5 625$

$\underline{\qquad 4 \qquad}$

Use the given *x*-values to graph each function. Then graph its inverse. Describe the domain and range of the inverse function.

13. $f(x) = 2^x$; $x = -2, -1, 0, 1, 2, 3, 4$

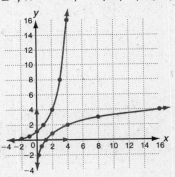

Domain: $\{x | x > 0\}$; range: all real numbers

14. $f(x) = \left(\frac{1}{2}\right)^x$; $x = -3, -2, -1, 0, 1, 2, 3$

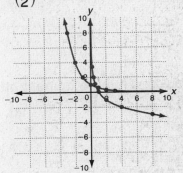

Domain: $\{x | x > 0\}$; range: all real numbers

Solve.

15. The hydrogen ion concentration in moles per liter for a certain brand of tomato-vegetable juice is 0.000316.

a. Write a logarithmic equation for the pH of the juice.

b. What is the pH of the juice?

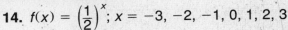

$\underline{pH = -\log (0.000316)}$

$\underline{\qquad 3.5 \qquad}$

Holt Algebra 2

LESSON 7-4 Practice B
Properties of Logarithms

Express as a single logarithm. Simplify, if possible.

1. $\log_3 9 + \log_3 27$

$$\log_3 243 = 5$$

2. $\log_2 8 + \log_2 16$

$$\log_2 128 = 7$$

3. $\log_{10} 80 + \log_{10} 125$

$$\log_{10} 10,000 = 4$$

4. $\log_6 8 + \log_6 27$

$$\log_6 216 = 3$$

5. $\log_3 6 + \log_3 13.5$

$$\log_3 81 = 4$$

6. $\log_4 32 + \log_4 128$

$$\log_4 4096 = 6$$

Express as a single logarithm. Simplify, if possible.

7. $\log_2 80 - \log_2 10$

$$\log_2 8 = 3$$

8. $\log_{10} 4000 - \log_{10} 40$

$$\log_{10} 100 = 2$$

9. $\log_4 384 - \log_4 6$

$$\log_4 64 = 3$$

10. $\log_2 1920 - \log_2 30$

$$\log_2 64 = 6$$

11. $\log_3 486 - \log_3 2$

$$\log_3 243 = 5$$

12. $\log_6 180 - \log_6 5$

$$\log_6 36 = 2$$

Simplify, if possible.

13. $\log_4 4^6$

$$6$$

14. $\log_5 5^{x-5}$

$$x - 5$$

15. $7^{\log_7 30}$

$$30$$

16. $12^{\log_{12} 1}$

$$1$$

17. $\log_8 8^5$

$$5$$

18. $\log_3 9^4$

$$8$$

Evaluate. Round to the nearest hundredth.

19. $\log_{12} 1$

$$0$$

20. $\log_3 30$

$$3.10$$

21. $\log_5 10$

$$1.43$$

Solve.

22. The Richter magnitude of an earthquake, M, is related to the energy released in ergs, E, by the formula $M = \frac{2}{3}\log\left(\frac{E}{10^{11.8}}\right)$. Find the energy released by an earthquake of magnitude 4.2.

$$10^{18.1} \text{ ergs}$$

Holt Algebra 2

LESSON 7-5 Practice B
Exponential and Logarithmic Equations and Inequalities

Solve and check.

1. $5^{2x} = 20$

$x \approx 0.9307$

2. $12^{2x-8} = 15$

$x \approx 4.5449$

3. $2^{x+6} = 4$

$x = -4$

4. $16^{5x} = 64^{x+7}$

$x = 3$

5. $243^{0.2x} = 81^{x+5}$

$x \approx -6.67$

6. $25^x = 125^{x-2}$

$x = 6$

7. $\left(\frac{1}{2}\right)^x = 16^2$

$x = -8$

8. $\left(\frac{1}{32}\right)^{2x} = 64$

$x = -0.6$

9. $\left(\frac{1}{27}\right)^{x-6} = 27$

$x = 5$

Solve.

10. $\log_4 x^5 = 20$

$x = 256$

11. $\log_3 x^6 = 12$

$x = 9$

12. $\log_4 (x-6)^3 = 6$

$x = 22$

13. $\log x - \log 10 = 14$

$x = 10^{15}$

14. $\log x + \log 5 = 2$

$x = 20$

15. $\log (x+9) = \log (2x-7)$

$x = 16$

16. $\log (x+4) - \log 6 = 1$

$x = 56$

17. $\log x^2 + \log 25 = 2$

$x = \pm 2$

18. $\log (x-1)^2 = \log (-5x-1)$

$x = -1, -2$

Use a table and graph to solve.

19. $2^{x-5} < 64$

$x < 11$

20. $\log x^3 = 12$

$x = 10,000$

21. $2^x 3^x = 1296$

$x = 4$

Solve.

22. The population of a small farming community is declining at a rate of 7% per year. The decline can be expressed by the exponential equation $P = C(1 - 0.07)^t$, where P is the population after t years and C is the current population. If the population was 8,500 in 2004, when will the population be less than 6,000?

2009

Holt Algebra 2

LESSON **Practice B**
7-6 **The Natural Base, e**

Graph.

1. $f(x) = e^{2x}$

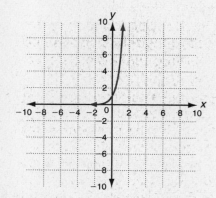

2. $f(x) = e^{0.5x}$

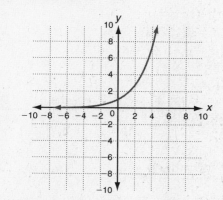

3. $f(x) = e^{1+x}$

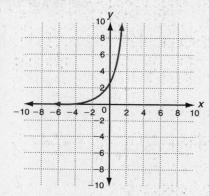

4. $f(x) = e^{2-x}$

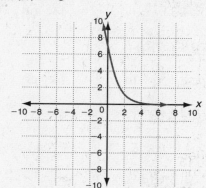

Simplify.

5. $\ln e^{x+2}$

$$x + 2$$

6. $e^{\ln 2x}$

$$2x$$

7. $e^{7\ln x}$

$$x^7$$

8. $\ln e^{3x+1}$

$$3x + 1$$

9. $\ln e$

$$1$$

10. $\ln e^{2x+y}$

$$2x + y$$

Solve.

11. Use the formula $A = Pe^{rt}$ to compute the total amount for an investment of $4500 at 5% interest compounded continuously for 6 years.

$$\$6074.36$$

12. Use the natural decay function, $N(t) = N_0 e^{-kt}$, to find the decay constant for a substance that has a half-life of 1000 years.

$$0.000693$$

Holt Algebra 2

Name _____ Date _____ Class _____

Practice B
Transforming Exponential and Logarithmic Functions

Graph each function. Find the asymptote. Tell how the graph is transformed from the graph of its parent function.

1. $f(x) = 5(2^x)$

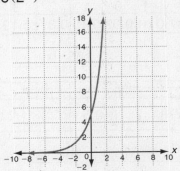

$y = 0$; it is the graph of $f(x) = 2^x$
stretched vertically by a factor of 5.

2. $f(x) = 5^{\frac{x}{4}}$

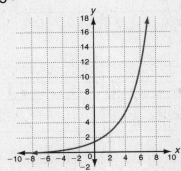

$y = 0$; it is the graph of $f(x) = 5^x$
stretched horizontally by a factor of 4.

3. $f(x) = \log(x + 5)$

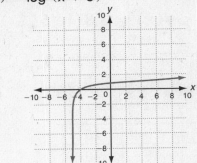

$x = -5$; it is the graph of $f(x) = \log$
x translated 5 units left.

4. $f(x) = 3 + \ln x$

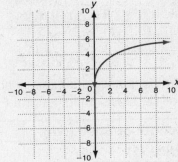

$x = 0$; it is the graph of $f(x) = \ln x$
translated 3 units up.

Write each transformed function.

5. The function $f(x) = \log(x + 1)$ is reflected across the y-axis
and translated down 4 units.

$$g(x) = \log(-x + 1) - 4$$

6. The function $f(x) = -8^{x-3}$ is reflected across the x-axis,
compressed horizontally by a factor of 0.2, and stretched
vertically by a factor of 2.

$$g(x) = 2 \cdot 8^{5x-3}$$

Solve.

7. The function $A(t) = Pe^{rt}$ can be used to calculate the growth of an investment
in which the interest is compounded continuously at an annual rate, r, over
t years. What annual rate would double an investment in 8 years?

8.7%

Holt Algebra 2

Name _____ Date _____ Class _____

Practice B
Curve Fitting with Exponential and Logarithmic Models

Determine whether *f* is an exponential function of *x*. If so, find the constant ratio.

1.

x	−1	0	1	2	3
f(x)	9	3	1	0.3	0.9

No

2.

x	−1	0	1	2	3
f(x)	0.01	0.03	0.15	0.87	5.19

Yes; 6

3.

x	−1	0	1	2	3
f(x)	$\frac{5}{6}$	$\frac{5}{2}$	7.5	22.5	67.5

Yes; 3

4.

x	−1	0	1	2	3
f(x)	1	0.5	0.33	0.25	0.2

No

Use exponential regression to find a function that models the data.

5.

x	1	2	3	4	5
f(x)	14	7.1	3.4	1.8	0.8

$$f(x) = 29(0.49)^x$$

6.

x	2	12	22	32	42
f(x)	5	20	80	320	1280

$$f(x) = 3.8(1.15)^x$$

Solve.

7. **a.** Bernice is selling seashells she has found at the beach. The price of each shell depends on its length. Find an exponential model for the data.

Length of Shell (cm)	5	8	12	20	25
Price ($)	2	3.5	5	18	40

$$f(x) = 0.97(1.16)^x$$

  b. What is the length of a shell selling for $9.00?

15 cm

  c. If Bernice found a 40 cm Conch shell. How much could she sell it for?

$367.36

8. **a.** Use logarithmic regression to find a function that models this data.

Time (min)	1	2	3	4	5
Speed (m/s)	1.5	6.2	10.6	12.9	14.8

$$f(x) = 1.14 + 8.42 \ln x$$

  b. When will the speed exceed 20 m/s?

9.4 s

  c. What will the speed be after 1 hour?

35.6 m/s

Holt Algebra 2

Name _____ Date _____ Class _____

Practice B
Variation Functions

Write and graph each function.

1. y varies directly as x, and $y = 30$ when $x = -6$.

$$y = -5x$$

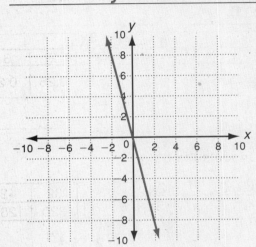

2. y varies inversely as x, and $y = 5$ when $x = 3$.

$$y = \frac{15}{x}$$

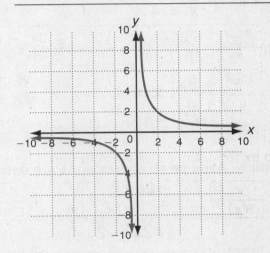

Determine whether each data set represents a direction variation, an inverse variation, or neither.

3.

x	8	12	16
y	2	3	4

In each case, $\frac{y}{x} = \frac{1}{4}$; the ratio is constant, so this represents a direct variation.

4.

x	3	1	0.5
y	5	15	30

In each case, $xy = 15$; the product is constant, so this represents an inverse variation.

Solve.

5. The number of chaperones, c, needed for the class trip varies directly as the number of students, s, going on the trip, and $c = 7$ when $s = 56$. How many chaperones are needed if 104 students go on the class trip?

13 chaperones

6. The owner of a bookstore developed a model for determining the price of rare comic books. The price, P, of each book should vary directly with the number of people, N, that have requested the book and inversely to the number of such books in existence, M. If $N = 10$ people, $M = 10,000$ copies and $P = \$5$, then find P for $N = 200$ people and $M = 100$ copies.

$10,000

Holt Algebra 2

Practice B

Multiplying and Dividing Rational Expressions

Simplify. Identify any *x*-values for which the expression is undefined.

1. $\dfrac{x^2 + 3x + 2}{x^2 - 3x - 4}$

$\dfrac{x + 2}{x - 4}; x \neq -1,$
$\quad x \neq 4$

2. $\dfrac{4x^6}{2x^4}$

$2x^2; x \neq 0$

3. $\dfrac{x^2 - x^3}{2x^2 - 5x + 3}$

$\dfrac{-x^2}{2x - 3}; x \neq 1,$
$\quad x \neq \dfrac{3}{2}$

4. $\dfrac{x^3 + x^2 - 20x}{x^2 - 16}$

$\dfrac{x^2 + 5x}{x + 4}; x \neq 4,$
$\quad x \neq -4$

5. $\dfrac{3x^2 - 9x - 12}{6x^2 + 9x + 3}$

$\dfrac{x - 4}{2x + 1}; x \neq -1,$
$\quad x \neq -\dfrac{1}{2}$

6. $\dfrac{9 - 3x}{15 - 2x - x^2}$

$\dfrac{3}{x + 5}; x \neq 3,$
$\quad x \neq -5$

Multiply. Assume all expressions are defined.

7. $\dfrac{4x + 16}{2x + 6} \cdot \dfrac{x^2 + 2x - 3}{x + 4}$

$2x - 2$

8. $\dfrac{x + 3}{x - 1} \cdot \dfrac{x^2 - 2x + 1}{x^2 + 5x + 6}$

$\dfrac{x - 1}{x + 2}$

Divide. Assume all expressions are defined.

9. $\dfrac{5x^6}{x^2y} \div \dfrac{10x^2}{y}$

$\dfrac{x^2}{2}$

10. $\dfrac{x^2 - 2x - 8}{x^2 - 2x - 15} \div \dfrac{2x^2 - 8x}{2x^2 - 10x}$

$\dfrac{x + 2}{x + 3}$

Solve. Check your solution.

11. $\dfrac{x^2 + x - 12}{x - 3} = 15$

$x = 11$

12. $\dfrac{2x^2 + 8x - 10}{2x^2 + 14x + 20} = 4$

$x = -3$

Solve.

13. The distance, *d*, traveled by a car undergoing constant acceleration, *a*, for a time, *t*, is given by $d = v_0 t + \dfrac{1}{2}at^2$, where v_0 is the initial velocity of the car. Two cars are side by side with the same initial velocity. One car accelerates, $a = A$, and the other car does not accelerate, $a = 0$. Write an expression for the ratio of the distance traveled by the accelerating car to the distance traveled by the nonaccelerating car as a function of time.

$1 + \dfrac{At}{2v_0}$

Holt Algebra 2

Practice B
LESSON 8-3
Adding and Subtracting Rational Expressions

Find the least common multiple for each pair.

1. $3x^2y^6$ and $5x^3y^2$

$$15x^3y^6$$

2. $x^2 + x - 2$ and $x^2 - x - 6$

$$(x - 1)(x + 2)(x - 3)$$

Add or subtract. Identify any *x*-values for which the expression is undefined.

3. $\dfrac{2x - 3}{x + 4} + \dfrac{4x - 5}{x + 4}$

$$\dfrac{6x - 8}{x + 4}; \; x \neq -4$$

4. $\dfrac{x + 12}{2x - 5} - \dfrac{3x - 2}{2x - 5}$

$$\dfrac{-2x + 14}{2x - 5}; \; x \neq \dfrac{5}{2}$$

5. $\dfrac{x + 4}{x^2 - x - 12} + \dfrac{2x}{x - 4}$

$$\dfrac{2x^2 + 7x + 4}{x^2 - x - 12}; \; x \neq 4, \; x \neq -3$$

6. $\dfrac{3x^2 - 1}{x^2 - 3x - 18} - \dfrac{x + 2}{x - 6}$

$$\dfrac{2x^2 - 5x - 7}{x^2 - 3x - 18}; \; x \neq 6, \; x \neq -3$$

7. $\dfrac{x + 2}{x^2 - 2x - 15} + \dfrac{x}{x + 3}$

$$\dfrac{x^2 - 4x + 2}{x^2 - 2x - 15}; \; x \neq -3, \; x \neq 5$$

8. $\dfrac{x + 6}{x^2 - 7x - 18} - \dfrac{2x}{x - 9}$

$$\dfrac{-2x^2 - 3x + 6}{x^2 - 7x - 18}; \; x \neq -2, \; x \neq 9$$

Simplify. Assume all expressions are defined.

9. $\dfrac{\frac{x - 1}{x + 5}}{\frac{x + 6}{x - 3}}$

$$\dfrac{x^2 - 4x + 3}{x^2 + 11x + 30}$$

10. $\dfrac{\frac{12}{x + 3}}{\frac{x^2 + 1}{x - 2}}$

$$\dfrac{12x - 24}{x^3 + 3x^2 + x + 3}$$

Solve.

11. A messenger is required to deliver 10 packages per day. Each day, the messenger works only for as long as it takes to deliver the daily quota of 10 packages. On average, the messenger is able to deliver 2 packages per hour on Saturday and 4 packages per hour on Sunday. What is the messenger's average delivery rate on the weekend?

$$2.6\overline{6} \text{ packages per hour}$$

Holt Algebra 2

Name _____ Date _____ Class _____

Practice B
Rational Functions

Using the graph of $f(x) = \frac{1}{x}$ as a guide, describe
the transformation and graph the function.

1. $g(x) = \frac{2}{x + 4}$

<u>**Translate 4 units left and
vertically stretched by a factor
of 2**</u>

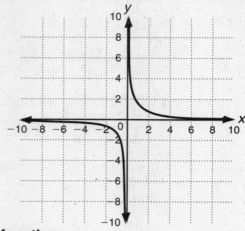

Identify the asymptotes, domain, and range of each function.

2. $g(x) = \frac{1}{x - 3} + 5$ <u>**Vertical asymptote: $x = 3$; horizontal asymptote: $y = 5$;
domain: $\{x \mid x \neq 3\}$; range: $\{y \mid y \neq 5\}$**</u>

3. $g(x) = \frac{1}{x + 8} - 1$ <u>**Vertical asymptote: $x = -8$; horizontal asymptote:
$y = -1$; domain: $\{x \mid x \neq -8\}$; range: $\{y \mid y \neq -1\}$**</u>

Identify the zeros and asymptotes of the function. Then graph.

4. $f(x) = \frac{x^2 + 4x - 5}{x + 1}$

a. Zeros:

<u>   **Zeros: -5 and 1**   </u>

b. Vertical asymptote:

<u> **Vertical asymptote: $x = -1$** </u>

c. Horizontal asymptote:

<u> **Horizontal asymptote: none** </u>

d. Graph.

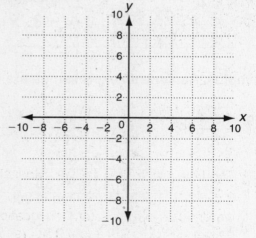

Solve.

5. The number, n, of daily visitors to a new store can be modeled by the function
$n = \frac{(250x + 1000)}{x}$, where x is the number of days the store has been open.

a. What is the asymptote of this function
and what does it represent?   <u>**The asymptote is 250; it is the average
number of people who will visit the store
each day, long after the store opens.**</u>

b. To the nearest integer, how many
visitors can be expected on day 30?   <u> **283** </u>

Holt Algebra 2

Practice B
Solving Rational Equations and Inequalities

Solve each equation.

1. $x - \dfrac{6}{x} = 5$

$x = -1$ or $x = 6$

2. $\dfrac{15}{4} = \dfrac{6}{x} + 3$

$x = 8$

3. $x = \dfrac{3}{x} + 2$

$x = 3$ or $x = -1$

4. $\dfrac{4}{x^2 - 4} = \dfrac{1}{x - 2}$

no solution.

Solve each inequality by using a graphing calculator and a table.

5. $\dfrac{6}{x + 1} < -3$

$-3 < x < -1$

6. $\dfrac{x}{x - 2} \geq 0$

$x \leq 0$ or $x > 2$

7. $\dfrac{2x}{x + 5} \leq 0$

$-5 < x \leq 0$

8. $\dfrac{-x}{x - 3} \geq 0$

$0 \leq x < 3$

Solve each inequality algebraically.

9. $\dfrac{12}{x + 4} \leq 4$

$x < -4$ or $x \geq -1$

10. $\dfrac{7}{x + 3} < -5$

$-\dfrac{22}{5} < x < -3$

11. $\dfrac{x}{x - 2} > 9$

$2 < x < \dfrac{9}{4}$

12. $\dfrac{2x}{x - 5} \geq 3$

$5 < x \leq 15$

Solve.

13. The time required to deliver and install a computer at a customer's location

is $t = 4 + \dfrac{d}{r}$, where t is time in hours, d is the distance, in miles, from the warehouse to the customer's location, and r is the average speed of the delivery truck. If it takes 6.2 hours for the employee to deliver and install a computer for a customer located 100 miles from the warehouse, what is the average speed of the delivery truck?

About 45.5 miles per hour

Holt Algebra 2

LESSON 8-6 Practice B
Radical Expressions and Rational Exponents

Simplify each expression. Assume all variables are positive.

1. $\sqrt[3]{125x^9}$

$5x^3$

2. $\sqrt[4]{\dfrac{x^8}{81}}$

$\dfrac{x^2}{3}$

3. $\sqrt[3]{\dfrac{64x^3}{8}}$

$2x$

Write each expression in radical form, and simplify.

4. $64^{\frac{5}{6}}$

32

5. $27^{\frac{2}{3}}$

9

6. $(-8)^{\frac{4}{3}}$

16

Write each expression by using rational exponents.

7. $\sqrt[5]{51^4}$

$51^{\frac{4}{5}}$

8. $(\sqrt{169})^3$

$169^{\frac{3}{2}}$

9. $\sqrt[7]{36^{14}}$

36^2

Simplify each expression.

10. $4^{\frac{3}{2}} \cdot 4^{\frac{5}{2}}$

256

11. $\dfrac{27^{\frac{4}{3}}}{27^{\frac{2}{3}}}$

9

12. $\left(125^{\frac{2}{3}}\right)^{\frac{1}{2}}$

5

13. $(27 \cdot 64)^{\frac{2}{3}}$

144

14. $\left(\dfrac{1}{243}\right)^{\frac{1}{5}}$

$\dfrac{1}{3}$

15. $64^{-\frac{1}{3}}$

$\dfrac{1}{4}$

16. $(-27x^6)^{\frac{1}{3}}$

$-3x^2$

17. $\dfrac{(25x)^{\frac{3}{2}}}{5 \cdot x^{\frac{1}{2}}}$

$25x$

18. $(4x)^{-\frac{1}{2}} \cdot (9x)^{\frac{1}{2}}$

$\dfrac{3}{2}$

Solve.

19. In every atom, electrons orbit the nucleus with a certain characteristic velocity known as the Fermi–Thomas velocity, equal to $\dfrac{Z^{\frac{2}{3}}}{137}$ c, where Z is the number of protons in the nucleus and c is the speed of light. In terms of c, what is the characteristic Fermi-Thomas velocity of the electrons in Uranium, for which $Z = 92$?

About $0.15c$

Holt Algebra 2

LESSON 8-7 Practice B
Radical Functions

Graph each function, and identify its domain and range.

1. $f(x) = \sqrt{x - 4}$

2. $f(x) = \sqrt[3]{x} + 1$

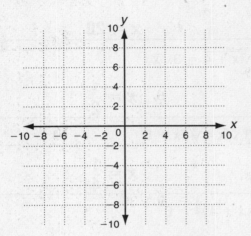

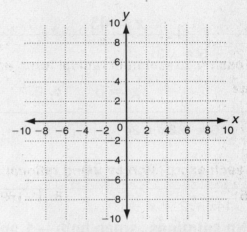

Domain: $\{x \mid x \geq -4\}$

Domain: **all real numbers**

Range: $\{y \mid y \geq 0\}$

Range: **all real numbers**

Using the graph of $f(x) = \sqrt{x}$ as a guide, describe the transformation.

3. $g(x) = 4\sqrt{x + 8}$ **Vertical stretch by a factor of 4 and translate 8 units left**

4. $g(x) = -\sqrt{3x} + 2$ **Reflection across the *x*-axis, horizontal compression by a factor of $\frac{1}{3}$, and translate 2 units up**

Use the description to write the square root function *g*.

5. The parent function $f(x) = \sqrt{x}$ is reflected across the *y*-axis, vertically stretched by a factor of 7, and translated 3 units down.

$$g(x) = 7\sqrt{-x} - 3$$

6. The parent function $f(x) = \sqrt{x}$ is translated 2 units right, compressed horizontally by a factor of $\frac{1}{2}$, and reflected across the *x*-axis.

$$g(x) = -\sqrt{2(x - 2)}$$

Solve.

7. For a gas with density, *n*, measured in atoms per cubic centimeter, the average distance, *d*, between atoms is given by $d = \left(\frac{3}{4\pi n}\right)^{\frac{1}{3}}$. The gas in a certain region of space has a density of just 10 atoms per cubic centimeter. Find the average distance between the atoms in that region of space.

0.29 cm

Holt Algebra 2

LESSON	**Practice B**
8-8	***Solving Radical Equations and Inequalities***

Solve each equation.

1. $\sqrt{x + 6} = 7$

$$x = 43$$

2. $\sqrt{5x} = 10$

$$x = 20$$

3. $\sqrt{2x + 5} = \sqrt{3x - 1}$

$$x = 6$$

4. $\sqrt{x + 4} = 3\sqrt{x}$

$$x = \frac{1}{2}$$

5. $\sqrt[3]{x - 6} = \sqrt[3]{3x + 24}$

$$x = -15$$

6. $3\sqrt[3]{x} = \sqrt[3]{7x + 5}$

$$x = \frac{1}{4}$$

7. $\sqrt{-14x + 2} = x - 3$

No solutions, since both −1 and −7 are extraneous

8. $(x + 4)^{\frac{1}{2}} = 6$

$$x = 32$$

9. $4(x - 3)^{\frac{1}{2}} = 8$

$$x = 7$$

10. $4(x - 12)^{\frac{1}{3}} = -16$

$$x = -52$$

Solve each inequality.

11. $\sqrt{3x + 6} \leq 3$

$$-2 \leq x \leq 1$$

12. $\sqrt{x - 4} + 3 > 9$

$$x > 40$$

13. $\sqrt{x + 7} \geq \sqrt{2x - 1}$

$$\frac{1}{2} \leq x \leq 8$$

14. $\sqrt{2x - 7} > 9$

$$x > 44$$

Solve.

15. A biologist is studying two species of animals in a habitat. The population, p_1, of one of the species is growing according to $p_1 = 500t^{\frac{3}{2}}$ and the population, p_2, of the other species is growing according to $p_2 = 100t^2$ where time, t, is measured in years. After how many years will the populations of the two species be equal?

25 years

Holt Algebra 2

Practice B
Multiple Representations of Functions

Match each situation to its corresponding graph. Sketch a possible graph of the situation if it does not match any of the given graphs.

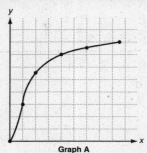
Graph A

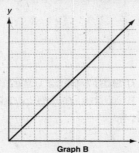
Graph B

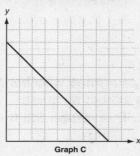
Graph C

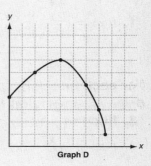
Graph D

1. A train is approaching its destination.

C

2. The temperature on an autumn day increases until late afternoon and then drops dramatically by late evening.

D

3. A helium balloon is released by a running child on a calm day.

B

4. A golf ball hit by a golfer flies over the trees and disappears into the woods.

A

Solve.

5. A bicyclist leaves a rest stop at 1:00 and heads directly for home at a constant rate. The table shows how far, d, he is from home in miles as a function of time, t. Create a graph and an equation to predict the time he will arrive home.

t	1:00	1:10	1:20	1:30	1:40
d	18.5	16.0	13.5	11.0	8.5

$$d = -0.25t + 18.5$$

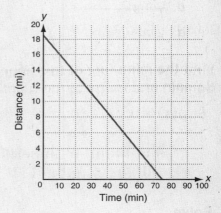

6. New members at a fitness club pay $200 to start and then $20 per month for life. Create a table, a graph, and an equation that represent the total cost of enrollment, c, as a function of months, m, of participation.

m	1	2	3
c	220	240	260

$$c = 20m + 200$$

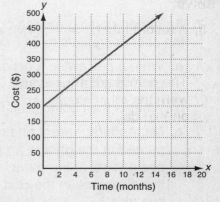

Holt Algebra 2

LESSON 9-2

Practice B
Piecewise Functions

Evaluate each piecewise function for $x = -8$ and $x = 5$.

1. $f(x) = \begin{cases} 2x & \text{if } x < 0 \\ 0 & \text{if } x \geq 0 \end{cases}$

_____ **−16, 0** _____

2. $g(x) = \begin{cases} 2 - x & \text{if } x \leq 5 \\ -x^2 & \text{if } 5 < x < 8 \\ 6 & \text{if } 8 \leq x \end{cases}$

_____ **10, −3** _____

3. $h(x) = \begin{cases} 2x + 4 & \text{if } x \leq -8 \\ -1 & \text{if } -8 < x < 5 \\ x^2 & \text{if } 5 \leq x \end{cases}$

_____ **−12, 25** _____

4. $k(x) = \begin{cases} 15 & \text{if } x \leq -5 \\ x & \text{if } -5 < x < 1 \\ 7 - \dfrac{x}{2} & \text{if } 1 < x \end{cases}$

15, $4\frac{1}{2}$

Graph each function.

5. $f(x) = \begin{cases} 6 & \text{if } x < -2 \\ 3x & \text{if } -2 \leq x \end{cases}$

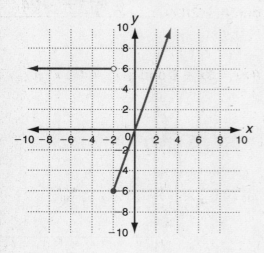

6. $g(x) = \begin{cases} 12 - x & \text{if } x \leq 5 \\ x + 2 & \text{if } 5 < x \end{cases}$

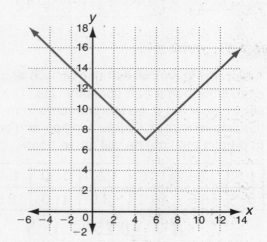

Solve.

7. An airport parking garage costs $20 per day for the first week. After that, the cost decreases to $17 per day.

a. Write a piecewise function for the cost of parking a car for x days.

$f(x) = \begin{cases} 20x & \text{if } x \leq 7 \\ 17x + 21 & \text{if } x > 7 \end{cases}$

b. What is the cost to park for 10 days?

$191

c. Ms. Anderson went on two trips. On the first, she parked at the garage for 5 days; on the second, she parked at the garage for 8 days. What was the difference in the cost of parking between the two trips?

$57

Holt Algebra 2

Name _____ Date _____ Class _____

Practice B
Transforming Functions

Given $f(x) = \begin{cases} x^2 - 9x - 1 & \text{if } x < 0 \\ 10 - x & \text{if } x \geq 0 \end{cases}$, write the rule for each function.

1. $h(x)$, a reflection of $f(x)$ across the y-axis

$h(x) = \begin{cases} -x^2 + 9x + 1 & \text{if } x < 0 \\ 10 + x & \text{if } x \geq 0 \end{cases}$

2. $k(x)$, a vertical stretch of $f(x)$ by a factor of 2

$k(x) = \begin{cases} 2x^2 - 18x - 2 & \text{if } x < 0 \\ 20 - 2x & \text{if } x \geq 0 \end{cases}$

3. $g(x)$, a horizontal translation 2 units right

$g(x) = \begin{cases} x^2 - 13x + 21 & \text{if } x < 2 \\ 12 - x & \text{if } x \geq 2 \end{cases}$

Identify the x- and y-intercepts of $f(x)$. Then identify the x- and y-intercepts of $g(x)$.

4. $f(x) = x^2 - 36$ ___ x-int. = 6 and -6, y-int. = -36 ___

$g(x) = f(2x)$ ___ x-int. = 3 and -3, y-int. = -36 ___

5. $f(x) = -3x + 12$ ___ x-int. = 4, y-int. = 12 ___

$g(x) = -2f(x)$ ___ x-int. = 4, y-int. = -24 ___

Given $f(x)$, graph $g(x)$.

6. $f(x) = x^2 + 2x + 1$ and $g(x) = -f\left(\dfrac{x}{2}\right)$

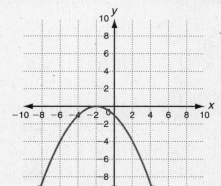

7. $f(x) = 3x - 6$ and $g(x) = f(-x)$

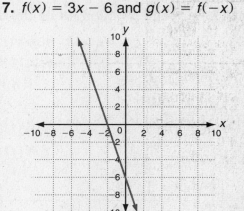

Solve.

8. Ron walks from his house to the parking garage at a rate of 8 feet per second. The parking garage is 3960 feet from the house. The distance can be represented by the function $D(x) = 8x$, where x is the time, in seconds. Walking back to his house, Ron increases his speed by 25%.

   a. Write a function to show the distance Ron is from the house as he walks back from the parking garage. ___ $D(x) = 3960 - 10x$ ___

   b. How far is Ron from his house 2 minutes after leaving the parking garage? ___ **2760 ft** ___

Holt Algebra 2

Practice B
Operations with Functions

Use the following functions for Exercises 1–18.

$$f(x) = \frac{1}{2x} \qquad g(x) = x^2 \qquad h(x) = x - 8 \qquad k(x) = \sqrt{x}$$

Find each function.

1. $(gk)(x)$

$$x^2\sqrt{x}$$

2. $(g + h)(x)$

$$x^2 + x - 8$$

3. $(g - h)(x)$

$$x^2 - x + 8$$

4. $(fg)(x)$

$$\frac{x}{2}$$

5. $(gh)(x)$

$$x^3 - 8x^2$$

6. $\left(\frac{f}{g}\right)(x)$

$$\frac{1}{2x^3}$$

Find each value.

7. $g(k(9))$

$$9$$

8. $h(g(-3))$

$$1$$

9. $g(h(-3))$

$$121$$

10. $k(h(12))$

$$2$$

11. $f(g(4))$

$$\frac{1}{32}$$

12. $f(h(1))$

$$-\frac{1}{14}$$

Write each composite function. State the domain of each.

13. $f(g(x))$

$$f(g(x)) = \frac{1}{2x^2};$$
$$\{x \mid x \neq 0\}$$

14. $h(g(x))$

$$h(g(x)) = x^2 - 8;$$
$$\{x \mid x \text{ is a real number}\}$$

15. $h(k(x))$

$$h(k(x)) = \sqrt{x} - 8;$$
$$\{x \mid x \geq 0\}$$

16. $f(k(x))$

$$f(k(x)) = \frac{\sqrt{x}}{2x};$$
$$\{x \mid x > 0\}$$

17. $k(g(x))$

$$k(g(x)) = \pm x;$$
$$\{x \mid x \text{ is a real number}\}$$

18. $k(h(x))$

$$k(h(x)) = \sqrt{x - 8};$$
$$\{x \mid x \geq 8\}$$

Solve.

19. A retail shoe store manager sets the price of shoes at twice his cost. The shoe store is now offering a 40% discount on all shoes.

a. Write a composite function for the price of a pair of shoes after the discount.

$$f(g(x)) = 1.2x$$

b. If a pair of shoes cost the manager $25, what is the sale price?

$$\$30$$

Holt Algebra 2

Practice B
Functions and Their Inverses

Find the inverse of each function. Determine whether the inverse is a function and state its domain and range.

1. $k(x) = 10x + 5$

$$k^{-1}(x) = \frac{x - 5}{10}; \text{ function}$$

domain: $(-\infty, +\infty)$

range: $(-\infty, +\infty)$

2. $d(x) = 6 - 2x$

$$d^{-1}(x) = -\frac{x}{2} + 3; \text{ function}$$

domain: $(-\infty, +\infty)$

range: $(-\infty, +\infty)$

3. $f(x) = (x - 5)^2$

$y = 5 \pm \sqrt{x}; \text{ not a function}$

domain: $(-\infty, +\infty)$

range: $[0, +\infty)$

4. $g(x) = \frac{4 - x}{2}$

$g^{-1}(x) = -2x + 4; \text{ function}$

domain: $(-\infty, +\infty)$

range: $(-\infty, +\infty)$

5. $h(x) = \sqrt{x^2 - 9}$

$h^{-1}(x) = \pm\sqrt{x^2 + 9}; \text{ not a function}$

domain: $[0, +\infty)$

range: $(-\infty, -3]$ and $[3, +\infty)$

6. $b(x) = 2\log x$

$b^{-1}(x) = \log^{-1}\frac{x}{2} \text{ or } b^{-1}(x) =$

$10^{\frac{x}{2}}; \text{ function; domain: } (-\infty, +\infty)$

range: $[0, +\infty)$

Determine by composition whether each pair of functions are inverses.

7. $q(x) = \sqrt{x} - 4$
and $r(x) = x^2 + 4$ for $x \geq 0$

No

8. $s(x) = \frac{2}{x - 2}$ and $t(x) = \frac{x + 2}{-2}$

No

9. $u(x) = \frac{x^2}{4} - 1$ for $x \geq -1$

and $v(x) = \pm 2\sqrt{x + 1}$

Yes

10. $A(x) = \log(x - 1)^4$

and $B(x) = 1 + \log^{-1}\left(\frac{x}{4}\right)$

Yes

Solve.

11. So far, Rhonda has saved $3000 for her college expenses. She plans to save $30 each month. Her college fund can be represented by the function $f(x) = 30x + 3000$.

   a. Find the inverse of $f(x)$.

   b. What does the inverse represent?

   c. When will the fund reach $3990?

   d. How long will it take her to reach her goal of $4800?

$$f^{-1}(x) = \frac{1}{30}x - 100$$

Number of months she has saved

33 months

5 years

Holt Algebra 2

Practice B
Modeling Real-World Data

Use constant differences or ratios to determine which parent function would best model the given data set.

1.

x	12	16	20	24	28
y	0.8	3.6	16.2	72.9	328.05

_____ **Exponential** _____

2.

x	13	19	25	31	37	43
y	−1	17	35	53	71	89

_____ **Linear** _____

3.

x	2	7	12	17	22
y	−100	−55	40	185	380

_____ **Quadratic** _____

4.

x	0.10	0.37	0.82	1.45	2.26
y	0.3	0.6	0.9	1.2	1.5

_____ **Square root** _____

Write a function that models the data set.

5.

x	2.2	2.6	3.0	3.4	3.8
y	0.68	4.52	9.0	14.12	19.88

$$f(x) = 2x^2 - 9$$

6.

x	−5	0	5	10	15	20
y	8	6	4	2	0	−2

$$f(x) = -0.4x + 6$$

7.

x	0.3	0.7	1.1	1.5	1.9
y	2.5	3	3.6	4.32	5.184

$$f(x) = 2.18(1.577)^x$$

8.

x	0.06	0.375	0.96	1.815	2.94
y	0.2	0.5	0.8	1.1	1.4

$$f(x) = 0.816\sqrt{x}$$

9.

x	−6	1	8	15	22
y	15	1	30.12	102.36	217.72

$$f(x) = 0.44x^2 + 0.2x + 0.36$$

10.

x	0.32	2.07	4.8	8.51	13.2
y	0.9	1.6	2.3	3.0	3.7

$$f(x) = 1.318x^{0.378}$$

Solve.

11. The table shows the population growth of a small town.

Years after 1974	1	6	11	16	21	26	31
Population	662	740	825	908	1003	1095	1200

a. Write a function that models the data. _____ $f(x) = 657.3(1.02)^x$

b. Use your model to predict the population in 2020. _____ **1634 people**

Holt Algebra 2

LESSON **10-1**

Practice B
Introduction to Conic Sections

Graph each equation on a graphing calculator. Identify each conic section. Then describe the center and intercepts.

1. $5x^2 + 5y^2 = 45$

    Circle; center: (0, 0); intercepts: (0, 3), (0, −3), (3, 0), (−3, 0)

2. $x^2 + 25y^2 = 25$

    Ellipse; center (0, 0); intercepts: (0, 1), (0, −1), (5, 0), (−5, 0)

3. $4x^2 + 4y^2 = 64$

    Circle; center (0, 0); intercepts: (0, 4), (0, −4), (4, 0), (−4, 0)

4. $49x^2 + 4y^2 = 196$

    Ellipse; center (0, 0); intercepts: (0, 7), (0, −7), (2, 0), (−2, 0)

Graph each equation on a graphing calculator. Identify each conic section. Then describe the vertices and the direction that the graph opens.

5. $y = x^2 − 3$

    Parabola; (0, −3); opens upward

6. $y^2 = x^2 + 4$

    Hyperbola; (0, 2), (0, −2); vertically

7. $x^2 − y^2 = 1$

    Hyperbola; (1, 0), (−1, 0); horizontally

8. $y^2 = x + 4$

    Parabola; (−4, 0); opens right

9. $x^2 = y + 2$

    Parabola; (0, −2); opens upward

10. $x = y^2 + 3$

    Parabola; (3, 0); opens right

Find the center and radius of a circle that has a diameter with the given endpoints.

11. (−10, 5) and (−5, 17)

    Center (−7.5, 11); radius = 6.5

12. (8, 1) and (−4, 10)

    Center (2, 5.5); radius = 7.5

13. (−7, 12) and (11, −68)

    Center (2, −28); radius = 41

Solve.

14. The orbit of an asteroid is modeled by the equation $9x^2 + 36y^2 = 144$.

    a. Identify the conic section.   Ellipse

    b. Identify the *x*- and *y*-intercepts of the orbit.

        (0, 2), (0, −2), (4, 0), (−4, 0)

    c. Suppose each unit of the coordinate plane represents 40 million miles. What is the maximum width of the asteroid's orbit?

        320 million miles

Holt Algebra 2

Practice B
Circles

Write the equation of each circle.

1. Center $(8, 9)$ and radius $r = 10$

$$(x - 8)^2 + (y - 9)^2 = 100$$

2. Center $(-1, 5)$ and containing the point $(23, -2)$

$$(x + 1)^2 + (y - 5)^2 = 625$$

3. Center $(2, 2)$ and containing the point $(-1, 6)$

$$(x - 2)^2 + (y - 2)^2 = 25$$

4. Center $(3, -5)$ and containing the point $(-7, 11)$

$$(x - 3)^2 + (y + 5)^2 = 356$$

5. Center $(-3, 0)$ and radius $r = 6$

$$(x + 3)^2 + y^2 = 36$$

6. Center $(6, -1)$ and radius $r = 8$

$$(x - 6)^2 + (y + 1)^2 = 64$$

7. Center $(-3, -4)$ and containing the point $(3, 4)$

$$(x + 3)^2 + (y + 4)^2 = 100$$

8. Center $(5, -5)$ and containing the point $(1, -2)$

$$(x - 5)^2 + (y + 5)^2 = 25$$

Write the equation of the line that is tangent to each circle at the given point.

9. $x^2 + y^2 = 169$; $(12, 5)$

$$y = -\frac{12}{5}x + \frac{169}{5}$$

10. $(x - 2)^2 + (y - 1)^2 = 25$; $(6, -2)$

$$y = \frac{4}{3}x - 10$$

11. $(x - 7)^2 + (y + 3)^2 = 625$; $(0, -21)$

$$y = \frac{-7}{18}x - 21$$

12. $(x + 3)^2 + (y + 6)^2 = 144$; $(-3, 6)$

$$y = 6$$

Solve.

13. A rock concert is located at the point $(-1, 1)$. The music can be heard up to 4 miles away. Use the equation of a circle to find the locations that are affected. Assume each unit of the coordinate plane represents 1 mile.

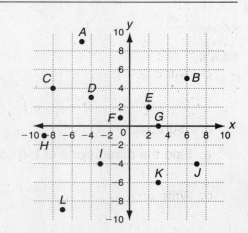

$$D, E, F$$

Holt Algebra 2

LESSON **Practice B**
10-3 *Ellipses*

Find the constant sum of an ellipse with the given foci and point on the ellipse.

1. $F_1(40, 0)$, $F_2(-40, 0)$, $P(0, -9)$

_____ **82** _____

2. $F_1(0, -20)$, $F_2(0, 20)$, $P(15, 0)$

_____ **50** _____

Write an equation in standard form for each ellipse with center $(0, 0)$.

3. Vertex $(15, 0)$, focus $(9, 0)$

$$\frac{x^2}{225} + \frac{y^2}{144} = 1$$

4. Co-vertex $(0, -21)$, focus $(-75, 0)$

$$\frac{x^2}{6066} + \frac{y^2}{441} = 1$$

5. Co-vertex $(-20, 0)$, focus $(0, 48)$

$$\frac{y^2}{2704} + \frac{x^2}{400} = 1$$

6. Vertex $(61, 0)$, focus $(60, 0)$

$$\frac{x^2}{3721} + \frac{y^2}{121} = 1$$

Graph each ellipse.

7. $\dfrac{(x+3)^2}{9} + \dfrac{(y-2)^2}{16} = 1$

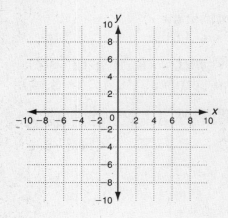

8. $\dfrac{(x-4)^2}{36} + \dfrac{(y-1)^2}{25} = 1$

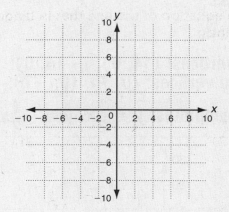

Solve.

9. Tom has a small semi-elliptical arch in his garden that he wants to enlarge. He wants to increase the height by a factor of 3 and increase the width by a factor of 2.5. The original arch can be modeled by the equation $\dfrac{y^2}{6.25} + \dfrac{x^2}{4} = 1$, measured in feet.

   a. Find the dimensions of the enlarged arch.

_____ **7.5 ft tall, 5 ft wide** _____

   b. Write an equation to model the enlarged arch.

$$\frac{y^2}{56.25} + \frac{x^2}{25} = 1$$

Holt Algebra 2

Name _____ Date _____ Class _____

Practice B
Hyperbolas

Find the constant difference for a hyperbola with the given foci and point on the hyperbola.

1. $F_1(0, 11)$, $F_2(0, -11)$, $P(0, 7)$

2. $F_1(-9, 0)$, $F_2(9, 0)$, $P(-8, 0)$

$$14$$

$$16$$

Write an equation in standard form for each hyperbola with center (0, 0).

3. Co-vertex $(-16, 0)$, focus $(0, -20)$

$$\frac{y^2}{144} - \frac{x^2}{256} = 1$$

4. Vertex $(24, 0)$, focus $(-25, 0)$

$$\frac{x^2}{576} - \frac{y^2}{49} = 1$$

5. Vertex $(0, -17)$, co-vertex $(1, 0)$

$$\frac{y^2}{289} - \frac{x^2}{1} = 1$$

6. Vertex $(30, 0)$, focus $(-40, 0)$

$$\frac{x^2}{900} - \frac{y^2}{700} = 1$$

Find the vertices, co-vertices, and asymptotes of each hyperbola, and then graph.

7. $\frac{x^2}{196} - \frac{y^2}{49} = 1$

Vertices: $(14, 0)$, $(-14, 0)$;
co-vertices: $(0, 7)$, $(0, -7)$;
asymptotes: $y = \frac{1}{2}x$, $y = -\frac{1}{2}x$

8. $\frac{(y-4)^2}{36} - \frac{x^2}{81} = 1$

Vertices: $(0, 10)$, $(0, -2)$;
co-vertices: $(9, 4)$, $(-9, 4)$;
asymptotes: $y = \frac{2}{3}x + 4$, $y = -\frac{2}{3}x + 4$

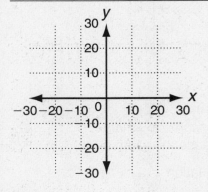

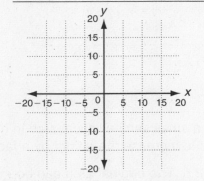

Solve.

9. A comet's path as it approaches the sun is modeled by one branch of the hyperbola $\frac{y^2}{1122} - \frac{x^2}{39,355} = 1$, where the sun is at the corresponding focus. Each unit of the coordinate plane represents one million miles. How close does the comet come to the sun?

$$\textbf{167.7 million miles}$$

74

Holt Algebra 2

LESSON 10-5 Practice B
Parabolas

Use the Distance Formula to find the equation of a parabola with the given focus and directrix.

1. $F(6, 0)$, $x = -3$

$$x = \frac{1}{18}y^2 + \frac{3}{2}$$

2. $F(1, 0)$, $x = -4$

$$x = 0.1y - 1.5$$

Write the equation in standard form for each parabola.

3. Vertex $(0, 0)$, directrix $y = -2$

$$y = \frac{1}{8}x^2$$

4. Vertex $(0, 0)$, focus $(9, 0)$

$$x = \frac{1}{36}y^2$$

5. Focus $(-6, 0)$, directrix $x = 6$

$$x = -\frac{1}{24}y^2$$

6. Vertex $(0, 0)$, focus $(0, -3)$

$$y = -\frac{1}{12}x^2$$

Find the vertex, value of p, axis of symmetry, focus, and directrix of each parabola. Then graph.

7. $x - 1 = -\frac{1}{12}y^2$

Vertex $(1, 0)$; $p = -3$; axis of symmetry $y = 0$; focus $(-2, 0)$; directrix $x = 4$

8. $y + 2 = \frac{1}{4}(x - 1)^2$

Vertex $(1, -2)$; $p = 1$; axis of symmetry $x = 1$; focus $(1, -1)$; directrix $y = -3$

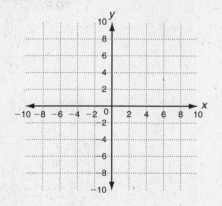

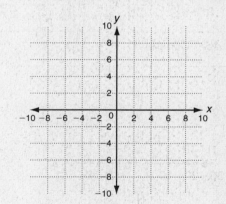

Solve.

9. A spotlight has parabolic cross sections.

a. Write an equation for a cross section of the spotlight if the bulb is 6 inches from the vertex and the vertex is placed at the origin.

$$y = \frac{1}{24}x^2$$

b. If the spotlight has a diameter of 36 inches at its opening, find the depth of the spotlight if the bulb is 6 inches from the vertex.

13.5 inches

Holt Algebra 2

Name _____ Date _____ Class _____

Identify the conic section that each equation represents.

1. $x - 1 = \frac{1}{4}(y - 8)^2$

   Parabola

2. $\frac{(y + 7)^2}{6^2} + \frac{(x - 9)^2}{1^2} = 1$

   Ellipse

3. $(x - 9)^2 + (y + 1)^2 = 3^2$

   Circle

4. $y + 5 = -(x - 9)^2$

   Parabola

5. $\frac{(y + 4)^2}{4^2} - \frac{(x - 4)^2}{3^2} = 1$

   Hyperbola

6. $\frac{(x - 2)^2}{6^2} + \frac{(y + 8)^2}{4^2} = 1$

   Ellipse

7. $y^2 + 8x + 2y + 57 = 0$

   Parabola

8. $x^2 + y^2 - 4x + 4y - 17 = 0$

   Circle

9. $x^2 - 9y^2 + 2x + 18y - 17 = 0$

   Hyperbola

10. $x^2 + 4y^2 - 2x - 16y + 1 = 0$

    Ellipse

Find the standard form of each equation by completing the square.
Then identify and graph each conic.

11. $x^2 - 16y^2 + 4x + 96y - 124 = 0$

$$\frac{(y - 3)^2}{1^2} - \frac{(x + 2)^2}{4^2} = 1;$$
hyperbola

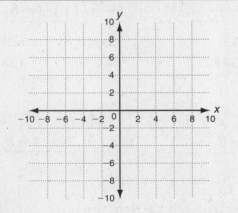

12. $x^2 + 4y^2 = 16$

$$\frac{x^2}{4^2} + \frac{y^2}{2^2} = 1; \text{ ellipse}$$

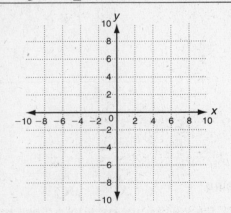

Solve.

13. A train takes a path around the town that can be modeled by the equation
$x^2 + 28x + 16y = 348$. The town lies at the focus.

   a. Write the equation in standard form.

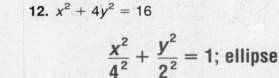
$$y - 34 = -\frac{1}{16}(x - 14)^2$$

   b. If the measurement is in miles, how close
   does the train come to the town?

   4 miles

Holt Algebra 2

Practice B

Solving Nonlinear Systems

Solve each system of equations by graphing.

1. $\begin{cases} 4x + y = 24 \\ x = \frac{1}{16}y^2 \end{cases}$

    $(9, -12), (4, 8)$

2. $\begin{cases} y - 4 = \frac{1}{4}x^2 \\ x + 2y = 12 \end{cases}$

    $(2, 5), (-4, 8)$

3. $\begin{cases} 9y - 6x = 0 \\ \frac{x^2}{45} + \frac{y^2}{5} = 1 \end{cases}$

    $(3, 2), (-3, -2)$

Solve each system of equations by using the substitution method.

4. $\begin{cases} x^2 + y^2 = 101 \\ 10x + y = 0 \end{cases}$

    $(1, -10), (-1, 10)$

5. $\begin{cases} 3y = 4x \\ x^2 - y^2 = -63 \end{cases}$

    $(9, 12), (-9, -12)$

6. $\begin{cases} 8y = x + 5 \\ x + 5 = \frac{1}{2}y^2 \end{cases}$

    $(-5, 0), (123, 16)$

7. $\begin{cases} x^2 + y^2 = 34 \\ 3x - 3y = 6 \end{cases}$

    $(5, 3), (-3, -5)$

8. $\begin{cases} x^2 + y^2 = 5 \\ y + 3 = \frac{1}{2}x^2 \end{cases}$

    $(2, -1), (-2, -1)$

9. $\begin{cases} x^2 + y^2 = 109 \\ x - 7 = \frac{1}{3}y^2 \end{cases}$

    $(10, 3), (10, -3)$

Solve each system of equations by using the elimination method.

10. $\begin{cases} 2x^2 + y^2 = 86 \\ x^2 + 3y^2 = 133 \end{cases}$

11. $\begin{cases} 4x^2 + y^2 = 13 \\ 2x^2 - y^2 = -7 \end{cases}$

12. $\begin{cases} 3x^2 + 2y^2 = 350 \\ 4x^2 - 2y^2 = -98 \end{cases}$

    $(5, 6), (5, -6),$
    $(-5, 6), (-5, -6)$

    $(1, 3), (1, -3),$
    $(-1, 3), (-1, -3)$

    $(6, 11), (6, -11),$
    $(-6, 11), (-6, -11)$

13. $\begin{cases} 8x^2 - 3y^2 = 173 \\ 5x^2 - y^2 = 116 \end{cases}$

14. $\begin{cases} 2x^2 - 3y^2 = 15 \\ 3x^2 + 2y^2 = 341 \end{cases}$

15. $\begin{cases} 5x^2 - 3y^2 = 128 \\ 4x^2 - 2y^2 = 128 \end{cases}$

    $(5, 3), (5, -3),$
    $(-5, 3), (-5, -3)$

    $(9, 7), (9, -7),$
    $(-9, 7), (-9, -7)$

    $(8, 8), (8, -8),$
    $(-8, 8), (-8, -8)$

Solve.

16. The shape of a state park can be modeled by a circle with the equation
    $x^2 + y^2 = 1600$. A stretch of highway near the park is modeled by the equation
    $y = \frac{1}{40}(x - 40)^2$. At what points does a car on the highway enter or exit the park?

    $(0, 40), (40, 0)$

Holt Algebra 2

Name _____ Date _____ Class _____

LESSON 11-1

Practice B
Permutations and Combinations

Use the Fundamental Counting Principle.

1. The soccer team is silk-screening T-shirts. They have 4 different colors of T-shirts and 2 different colors of ink. How many different T-shirts can be made using one ink color on a T-shirt?

   8 T-shirts

2. A travel agent is offering a vacation package. Participants choose the type of tour, a meal plan, and a hotel class from the table below.

Tour	Meal	Hotel
Walking	Restaurant	4-Star
Boat	Picnic	3-Star
Bicycle		2-Star
		1-Star

   How many different vacation packages are offered?

   24 packages

Evaluate.

3. $\frac{3!6!}{3!}$

   720

4. $\frac{10!}{7!}$

   720

5. $\frac{9! - 6!}{(9 - 6)!}$

   60,360

Solve.

6. In how many ways can the debate team choose a president and a secretary if there are 10 people on the team?

   90 ways

7. A teacher is passing out first-, second-, and third-place prizes for the best student actor in a production of *Hamlet*. If there are 14 students in the class, in how many different ways can the awards be presented?

   2184 ways

Evaluate.

8. $_5P_4$

   120

9. $_3C_2$

   3

10. $_8P_3$

   336

Solve.

11. Mrs. Marshall has 11 boys and 14 girls in her kindergarten class this year.

    a. In how many ways can she select 2 girls to pass out a snack?

       91 ways

    b. In how many ways can she select 5 boys to pass out new books?

       462 ways

    c. In how many ways can she select 3 students to carry papers to the office?

       2300 ways

Holt Algebra 2

Practice B
LESSON
11-2 *Theoretical and Experimental Probability*

Solve.

1. A fruit bowl contains 4 green apples and 7 red apples. What is the probability that a randomly selected apple will be green? _____ $\dfrac{4}{11}$

2. When two number cubes labeled 1–6 are rolled, what is the probability that the result will be two 4's? _____ $\dfrac{1}{36}$

3. Joanne is guessing which day in November is Bess's birthday. Joanne knows that Bess's birthday does not fall on an odd-numbered day. What is the probability that Joanne will guess the correct day on her first try? _____ $\dfrac{1}{15}$

4. Tom has a dollar's worth of dimes and a dollar's worth of nickels in his pocket.

   a. What is the probability he will randomly select a nickel from his pocket? _____ $\dfrac{2}{3}$

   b. What is the probability he will randomly select a dime from his pocket? _____ $\dfrac{1}{3}$

5. Clarice has 7 new CDs; 3 are classical music and the rest are pop music. If she randomly grabs 3 CDs to listen to in the car on her way to school, what is the probability that she will select only classical music? _____ $\dfrac{1}{35}$

6. Find the probability that a point chosen at random inside the larger circle shown here will also fall inside the smaller circle.

   $\dfrac{9}{16}$ _____

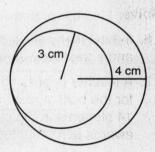

Frank is playing darts. The results of his throws are shown in the table below. Assume that his results continue to follow this trend.

Color Hit	Number of Throws
Blue	12
Red	5
White	2

Find the experimental probability of each event.

7. Frank's next throw will hit white. _____ $\dfrac{2}{19}$

8. Frank's next throw will hit blue. _____ $\dfrac{12}{19}$

9. Frank's next throw will hit either red or white. _____ $\dfrac{7}{19}$

10. Frank's next throw will NOT hit red. _____ $\dfrac{14}{19}$

 Holt Algebra 2

Practice B
Independent and Dependent Events

Find each probability.

1. A bag contains 5 red, 3 green, 4 blue, and 8 yellow marbles. Find the probability of randomly selecting a green marble, and then a yellow marble if the first marble is replaced. $\dfrac{3}{50}$

2. A sock drawer contains 5 rolled-up pairs of each color of socks, white, green, and blue. What is the probability of randomly selecting a pair of blue socks, replacing it, and then randomly selecting a pair of white socks? $\dfrac{1}{9}$

Two 1–6 number cubes are rolled—one is black and one is white.

3. The sum of the rolls is greater than or equal to 6 and the black cube shows a 3.

   a. Explain why the events are dependent. **The events are dependent because** $P(\text{sum} \geq 6)$ **is different when it is known that a black 3 occurred.**

   b. Find the probability. $\dfrac{1}{9}$

4. The white cube shows an even number, and the sum is 8.

   a. Explain why the events are dependent. **The events are dependent because** $P(\text{sum} = 8)$ **is different when it is known that the white cube shows an even number.**

   b. Find the probability. $\dfrac{1}{12}$

The table below shows numbers of registered voters by age in the United States in 2004 based on the census. Find each probability in decimal form.

Age	Registered Voters (in thousands)	Not Registered to Vote (in thousands)
18–24	14,334	13,474
25–44	49,371	32,763
45–64	51,659	19,355
65 and over	26,706	8,033

5. A randomly selected person is registered to vote, given that the person is between the ages of 18 and 24.   **0.52**

6. A randomly selected person is between the ages of 45 and 64 and is not registered to vote.   **0.09**

7. A randomly selected person is registered to vote and is at least 65 years old.   **0.12**

A bag contains 12 blue cubes, 12 red cubes, and 20 green cubes. Determine whether the events are independent or dependent, and find each probability.

8. A green cube and then a blue cube are chosen at random with replacement.   **Independent;** $\dfrac{15}{121}$

9. Two blue cubes are chosen at random without replacement.   **Dependent;** $\dfrac{3}{43}$

Holt Algebra 2

Name _____ Date _____ Class _____

A can of vegetables with no label has a $\frac{1}{8}$ chance of being green
beans and a $\frac{1}{5}$ chance of being corn.

1. Explain why the events "green beans" or "corn" are mutually exclusive.

   These events are mutually exclusive because each can contains only one type of vegetable.

2. What is the probability that an unlabeled can of vegetables $\frac{13}{40}$
   is either green beans or corn?

Ben rolls a 1–6 number cube. Find each probability.

3. Ben rolls a 3 or a 4. $\frac{1}{3}$

4. Ben rolls a number greater than 2 or an even number. $\frac{5}{6}$

5. Ben rolls a prime number or an odd number. $\frac{2}{3}$

Of the 400 doctors who attended a conference, 240 practiced family
medicine and 130 were from countries outside the United States.
One-third of the family medicine practitioners were not from the
United States.

6. What is the probability that a doctor practices family $\frac{7}{8}$
   medicine or is from the United States?

7. What is the probability that a doctor practices family $\frac{29}{40}$
   medicine or is not from the United States?

8. What is the probability that a doctor does not practice $\frac{4}{5}$
   family medicine or is from the United States?

Use the data to fill in the Venn diagram. Then solve.

9. Of the 220 people who came into the Italian deli on Friday, 104 bought
   pizza and 82 used a credit card. Half of the people who bought pizza
   used a credit card. What is the probability that a customer bought pizza
   or used a credit card?

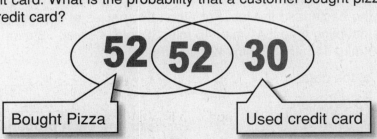

Bought Pizza   Used credit card

$\frac{67}{110}$ or 0.61

Solve.

10. There are 6 people in a gardening club. Each gardener orders seeds
    from a list of 11 different types of seeds available. What is the probability
    that 2 gardeners will order the same type of seeds?   0.81

81   **Holt Algebra 2**

Practice B
Measures of Central Tendency and Variation

Find the mean, median, and mode of each data set.

1. {12, 11, 17, 3, 9, 14, 16, 2}

    a. Mean _____ **10.5** _____

    b. Median _____ **11.5** _____

    c. Mode _____ **None** _____

2. {6, 9, 9, 20, 4, 5, 9, 13, 10, 1}

    a. Mean _____ **8.6** _____

    b. Median _____ **9** _____

    c. Mode _____ **9** _____

Make a box-and-whisker plot of the data. Find the interquartile range.

3. {3, 7, 7, 3, 10, 1, 6, 6}

    Interquartile range is 4.

4. {1, 2, 3, 5, 3, 5, 8, 2}

    Interquartile range is 3.

Find the variance and standard deviation.

5. {7, 4, 3, 9, 2}

    6.8; 2.6

6. {35, 67, 21, 16, 24, 51, 18, 32}

    278; 16.7

7. {19, 23, 17, 20, 25, 19, 15, 22}

    9.3; 3.0

8. {5, 12, 10, 13, 8, 11, 15, 12}

    8.4; 2.9

Solve.

9. The probability distribution for the amount of rain that falls on Boston in May each year is given below. Find the expected amount of rain for Boston in May.   **7.01**

Inches of Rain, *n*	5	6	7	8
Probability	0.05	0.10	0.64	0.21

10. A biologist is growing bacteria in the lab. For a certain species of bacteria, she records these doubling times: 41 min, 45 min, 39 min, 42 min, 38 min, 88 min, 43 min, 40 min, 44 min, 39 min, 42 min, and 40 min.

    a. Find the mean of the data.   **45.1**

    b. Find the standard deviation.   **13.1**

    c. Identify any outliers.   **88**

    d. Describe how any outlier affects the mean and the standard deviation.
    The mean increases from $\approx$41.2 to $\approx$45.1, and the standard deviation increases from $\approx$2.1 to $\approx$13.1.

Holt Algebra 2

LESSON **Practice B**
11-6 *Binomial Distributions*

Use the Binomial Theorem to expand each binomial.

1. $(x + y)^3$

$$x^3 + 3x^2y + 3xy^2 + y^3$$

2. $(2x + y)^4$

$$16x^4 + 32x^3y + 24x^2y^2 + 8xy^3 + y^4$$

3. $(m + 3n)^3$

$$m^3 + 9m^2n + 27mn^2 + 27n^3$$

4. $(p + q)^5$

$$p^5 + 5p^4q + 10p^3q^2 + 10p^2q^3 + 5pq^4 + q^5$$

Solve.

5. Of the new cars in a car dealer's lot, 1 in 6 are white. Today, 4 cars were sold.

   a. What is the probability that 3 of the cars sold were white?

   0.015

   b. What is the probability that at least 2 of the cars sold were white?

   0.13

6. At a small college, $\frac{1}{3}$ of all of the students are vegetarians. There are 5 students in line at the cafeteria.

   a. What is the probability that all 5 students are vegetarians?

   0.004

   b. What is the probability that just 1 of the students is a vegetarian?

   0.33

7. Ellen plays 8 hands of a card game with her friends. She has a 1 in 3 chance of winning each hand. What is the probability that she will win exactly half of the hands played?

   0.17

8. In a lottery, each ticket buyer has a 1 in 10 chance of winning a prize. If Chip buys 10 tickets, what is the probability that he will win at least 1 prize?

   0.65

Holt Algebra 2

LESSON
12-1

Practice B

Introduction to Sequences

Find the first 5 terms of each sequence.

1. $a_1 = 1$, $a_n = 3(a_{n-1})$

$$\underline{1, 3, 9, 27, 81}$$

2. $a_1 = 2$, $a_n = 2(a_{n-1} + 1) - 5$

$$\underline{2, 1, -1, -5, -13}$$

3. $a_1 = -2$, $a_n = (a_{n-1})^2 - 1$

$$\underline{-2, 3, 8, 63, 3968}$$

4. $a_1 = 1$, $a_n = 6 - 2(a_{n-1})$

$$\underline{1, 4, -2, 10, -14}$$

5. $a_1 = -1$, $a_n = (a_{n-1} - 1)^2 - 3$

$$\underline{-1, 1, -3, 13, 141}$$

6. $a_1 = -2$, $a_n = \dfrac{2 - a_{n-1}}{2}$

$$\underline{-2, 2, 0, 1, \tfrac{1}{2}}$$

7. $a_n = (n - 2)(n + 1)$

$$\underline{-2, 0, 4, 10, 18}$$

8. $a_n = n(2n - 1)$

$$\underline{1, 6, 15, 28, 45}$$

9. $a_n = n^3 - n^2$

$$\underline{0, 4, 18, 48, 100}$$

10. $a_n = \left(\dfrac{1}{2}\right)^{n-3}$

$$\underline{4, 2, 1, \tfrac{1}{2}, \tfrac{1}{4}}$$

11. $a_n = (-2)^{n-1}$

$$\underline{1, -2, 4, -8, 16}$$

12. $a_n = n^2 - 2n$

$$\underline{-1, 0, 3, 8, 15}$$

Write a possible explicit rule for the *n*th term of each sequence.

13. 8, 16, 24, 32, 40, …

$$\underline{a_n = 8n}$$

14. 0.1, 0.4, 0.9, 1.6, 2.5, …

$$\underline{a_n = 0.1n^2}$$

15. 3, 6, 11, 18, 27, …

$$\underline{a_n = n^2 + 2}$$

16. $\dfrac{3}{2}, \dfrac{3}{4}, \dfrac{3}{8}, \dfrac{3}{16}, \dfrac{3}{32}, \dots$

$$\underline{a_n = 3\left(\dfrac{1}{2}\right)^n}$$

17. −2, 1, 4, 7, 10, …

$$\underline{a_n = 3n - 5}$$

18. 5, 1, 0.2, 0.04, 0.008, …

$$\underline{a_n = 5(0.2)^{n-1}}$$

Solve.

19. Find the number of line segments in the next two iterations. __**31, 63**__

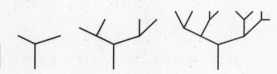

20. Jim charges \$50 per week for lawn mowing and weeding services. He plans to increase his prices by 4% each year.

a. Graph the sequence.

b. Describe the pattern.

__**Exponential**__

c. To the nearest dollar, how much will he charge per week in 5 years?

__**\$61 per week**__

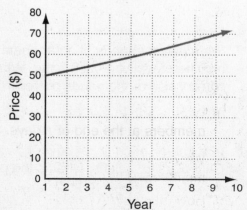

Holt Algebra 2

Practice B

LESSON 12-2 *Series and Summation Notation*

Write each series in summation notation.

1. $-2 + 4 - 8 + 16 - 32$

$$\sum_{k=1}^{5} (-2)^k$$

2. $\frac{1}{10} + \frac{1}{100} + \frac{1}{1,000} + \frac{1}{10,000}$

$$\sum_{k=1}^{4} \left(\frac{1}{10}\right)^k$$

3. $-6 - 1 + 4 + 9 + 14 + 19$

$$\sum_{k=1}^{6} (5k - 11)$$

4. $\frac{1}{3} + \frac{1}{6} + \frac{1}{9} + \frac{1}{12} + \frac{1}{15} + \frac{1}{18}$

$$\sum_{k=1}^{6} \frac{1}{3k}$$

5. $7 + 13 + 19 + 25 + 31$

$$\sum_{k=1}^{5} (6k + 1)$$

6. $-1 + 1 - 1 + 1 - 1 + 1 - 1$

$$\sum_{k=1}^{7} (-1)^k$$

Expand each series and evaluate.

7. $\sum_{k=4}^{8} \frac{k}{4}$

a. Expand. $\quad 1 + \frac{5}{4} + \frac{6}{4} + \frac{7}{4} + 2$

b. Simplify. $\quad 7\frac{1}{2}$

8. $\sum_{k=1}^{4} 5^{k-2}$

a. Expand. $\quad \frac{1}{5} + 1 + 5 + 25$

b. Simplify. $\quad 31\frac{1}{5}$

9. $\sum_{k=2}^{6} (-2^k)$

a. Expand. $\quad 4 - 8 + 16 - 32 + 64$

b. Simplify. $\quad 44$

10. $\sum_{k=30}^{39} (70 - 2k)$

a. Expand. $\quad 10 + 8 + 6 + 4 + 2 + 0 - 2 - 4 - 6 - 8$

b. Simplify. $\quad 10$

Evaluate each series.

11. $\sum_{k=12}^{20} 3$

27

12. $\sum_{k=1}^{40} k$

820

13. $\sum_{k=1}^{10} k^2$

385

Solve.

14. One day, Hannah starts a new online Internet club by convincing two of her friends to join. The next day, each member convinces two more people to join. The third day of the club, each member convinces two more people to join, and so on for a full week.

a. Write a series that represents the number of club members at the end of *n* days.

$$\sum_{k=1}^{n} 3^k$$

b. Write a series that represents the number of club members at the end of one week.

$$\sum_{k=1}^{7} 3^k$$

c. How many members will the club have at the end of a week?

3279

Holt Algebra 2

Practice B

Arithmetic Sequences and Series

Determine whether each sequence could be arithmetic. If so, find the common difference and the next term.

1. 41, 24, 7, −10, −27, …

$$-17; -44$$

2. 6, −6, 6, −6, 6, −6, 6, −6, …

Not arithmetic

3. $\frac{4}{5}, \frac{13}{10}, \frac{9}{5}, \frac{23}{10}, \frac{14}{5}, …$

$$\frac{1}{2}, \frac{33}{10}$$

4. 2, 4, 8, 16, 32, 64, …

Not arithmetic

Find the 12th term of each arithmetic sequence.

5. 21, 32, 43, 54, 65, …

$$142$$

6. 3.7, 3.3, 2.9, 2.5, 2.1, …

$$-0.7$$

7. 1.8, −1.1, −4, −6.9, −9.8, …

$$-30.1$$

8. −8, −2.75, 2.5, 7.75, 13, …

$$49.75$$

Find the missing terms in each arithmetic sequence.

9. 3, _, _, _, 59, …

$$17, 31, 45$$

10. −4, _, _, 23, …

$$5, 14$$

11. 7, _, _, _, _, 62, …

$$18, 29, 40, 51$$

12. 35, _, _, _, _, _, _, −7, …

$$29, 23, 17, 11, 5, -1$$

Find the 10th term of each arithmetic sequence.

13. $a_4 = 12$ and $a_7 = 20.4$

$$28.8$$

14. $a_3 = 37$ and $a_{17} = -12$

$$12.5$$

15. $a_{13} = -5$ and $a_{18} = -51$

$$22.6$$

16. $a_{25} = 18$ and $a_{41} = 62$

$$-23.25$$

Solve.

17. A banquet hall uses tables that seat 4, one person on each side. For a large party, the tables are positioned end to end in a long row. Two tables will seat 6, three tables will seat 8, and four tables will seat 10. How many tables should be set end to end to seat 40?

$$\text{19 tables}$$

Holt Algebra 2

Name _____ Date _____ Class _____

Determine whether each sequence could be geometric or arithmetic.
If possible, find the common ratio or difference.

1. 1.1, −3.3, 9.9, −29.7, 89.1, …   **2.** −18, −7, 4, 15, 26, …

Geometric; $r = -3$ Arithmetic; $d = 11$

3. 1, 2, 6, 24, 120, 720, …   **4.** 3125, 2500, 2000, 1600, 1280, …

Neither   Geometric; $r = 0.8$

Find the 10th term of each geometric sequence.

5. 1600, 800, 400, 200, …   **6.** 0.0000001, 0.00001, 0.001, 0.1, …

3.125   100,000,000,000

7. −64, 96, −144, 216, …   **8.** 2, −6, 18, −54, …

2460.375   −39,366

Find the 8th term of each geometric sequence with the given terms.

9. $a_3 = 12$ and $a_6 = 96$ **10.** $a_{15} = 100$ and $a_{17} = 25$

384   ±12,800

11. $a_{11} = -4$ and $a_{13} = -36$ **12.** $a_3 = -4$ and $a_5 = -36$

$\pm \dfrac{4}{27}$ ±972

Find the geometric mean of each pair of numbers.

13. 2 and 8   **14.** 4 and 25   **15.** 2 and 3

±4   ±10 $\pm\sqrt{6}$

Find the indicated sum for each geometric series.

16. S_7 for 14, 42, 126, 378, …   **17.** $\displaystyle\sum_{k=1}^{8} (-4)^{k-1}$

15,302   −13,107

Solve.

18. Deanna received an e-mail asking her to forward it to 10 other people.
Assume that no one breaks the chain and that there are no duplicate recipients.
How many e-mails will have been sent after 8 generations, including Deanna's?

111,111,111

                   Holt Algebra 2

Name _____ Date _____ Class _____

Practice B
Mathematical Induction and Infinite Geometric Series

Determine whether each geometric series converges or diverges.

1. $\frac{81}{625} + \frac{27}{125} + \frac{9}{25} + \frac{3}{5} + 1 + \cdots$

   Diverges

2. $1 - \frac{3}{5} + \frac{9}{25} - \frac{27}{125} + \frac{81}{625} - \cdots$

   Converges

Find the sum of each infinite geometric series, if it exists.

3. $7 + \frac{7}{4} + \frac{7}{16} + \frac{7}{64} + \cdots$

   $\frac{28}{3}$

4. $500 - 300 + 180 - 108 + \cdots$

   312.5

5. $\sum_{k=1}^{\infty} \frac{1}{4}\left(\frac{4}{3}\right)^k$

   Does not exist

6. $\sum_{k=1}^{\infty} 99\left(-\frac{4}{9}\right)^k$

   ≈ -30.46

Write each repeating decimal as a fraction in simplest form.

7. $0.\overline{16}$

   $\frac{16}{99}$

8. $0.\overline{016}$

   $\frac{16}{999}$

9. $0.0\overline{16}$

   $\frac{16}{990}$

10. $0.0\overline{45}$

    $\frac{1}{22}$

11. $0.\overline{1}$

    $\frac{1}{9}$

12. $0.\overline{123}$

    $\frac{41}{333}$

Identify a counterexample to disprove each statement.

13. $2^{-n} < n^2$

    Possible answer: $n = -5$

14. $n^3 \geq 3n$

    Possible answer: $n = -2$

Solve.

15. Ron won a prize that pays $200,000 the first year and half of the previous year's amount each year for the rest of his life.

   a. Write the first 4 terms of a series to represent the situation.

   $$200,000 + 100,000 + 50,000 + 25,000 + \cdots$$

   b. Write a general rule for a geometric sequence that models his prize each year.

   $a_n = 200,000(0.5)^{n-1}$

   c. Estimate Ron's total prize in the first 10 years.

   $\approx \$399,609.38$

   d. If Ron lives forever, what is the total of his winnings?

   $\$400,000$

Holt Algebra 2

Name _____ Date _____ Class _____

Practice B
Right-Angle Trigonometry

Find the value of the sine, cosine, and tangent functions for θ.

1.

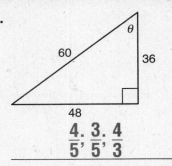

$$\frac{4}{5}, \frac{3}{5}, \frac{4}{3}$$

2.

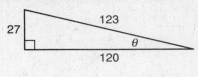

$$\frac{9}{41}, \frac{40}{41}, \frac{9}{40}$$

3.

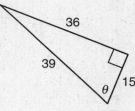

$$\frac{12}{13}, \frac{5}{13}, \frac{12}{5}$$

Use a trigonometric function to find the value of x.

4.

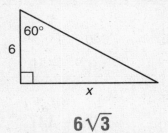

$$6\sqrt{3}$$

5.

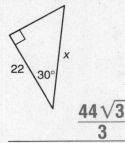

$$\frac{44\sqrt{3}}{3}$$

6.

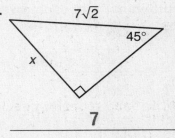

$$7$$

Find the values of the six trigonometric functions for θ.

7.

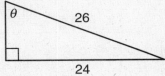

$$\sin\theta = \frac{12}{13}; \cos\theta = \frac{5}{13}; \tan\theta = \frac{12}{5}$$

$$\csc\theta = \frac{13}{12}; \sec\theta = \frac{13}{5}; \cot\theta = \frac{5}{12}$$

8.

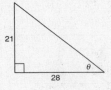

$$\sin\theta = \frac{3}{5}; \cos\theta = \frac{4}{5}; \tan\theta = \frac{3}{4}$$

$$\csc\theta = \frac{5}{3}; \sec\theta = \frac{5}{4}; \cot\theta = \frac{4}{3}$$

9.

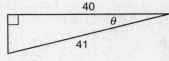

$$\sin\theta = \frac{9}{41}; \cos\theta = \frac{40}{41}; \tan\theta = \frac{9}{40}$$

$$\csc\theta = \frac{41}{9}; \sec\theta = \frac{41}{40}; \cot\theta = \frac{40}{9}$$

Solve.

10. A water slide is 26 feet high. The angle between the slide and the water is 33.5°. What is the length of the slide?

_____ **47 ft** _____

11. A surveyor stands 150 feet from the base of a viaduct and measures the angle of elevation to be 46.2°. His eye level is 6 feet above the ground. What is the height of the viaduct to the nearest foot?

_____ **162 ft** _____

12. The pilot of a helicopter measures the angle of depression to a landing spot to be 18.8°. If the pilot's altitude is 1640 meters, what is the horizontal distance to the landing spot to the nearest meter?

_____ **4817 m** _____

Holt Algebra 2

LESSON 13-2 Practice B
Angles of Rotation

Draw an angle with the given measure in standard position.

1. −420°

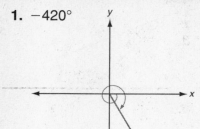

2. 405°

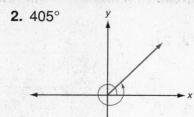

3. −450°

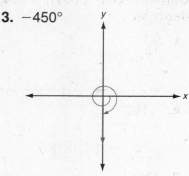

Find the measures of a positive angle and a negative angle that are coterminal with each given angle.

4. $\theta = 425°$

$65°, -295°$

5. $\theta = -316°$

$44°, -676°$

6. $\theta = -800°$

$280°, -80°$

7. $\theta = 281°$

$641°, -79°$

8. $\theta = -4°$

$356°, -364°$

9. $\theta = 743°$

$23°, -337°$

Find the measure of the reference angle for each given angle.

10. $\theta = 211°$

$31°$

11. $\theta = -755°$

$35°$

12. $\theta = -555°$

$15°$

13. $\theta = 119°$

$61°$

14. $\theta = -160°$

$20°$

15. $\theta = 235°$

$55°$

P is a point on the terminal side of θ in standard position. Find the exact value of the six trigonometric functions for θ.

16. $P(-5, 5)$

$\sin\theta = \dfrac{\sqrt{2}}{2}; \cos\theta = -\dfrac{\sqrt{2}}{2};$

$\tan\theta = -1; \cot\theta = -1;$

$\csc\theta = \sqrt{2}; \sec\theta = -\sqrt{2}$

17. $P(2, 9)$

$\sin\theta = \dfrac{9\sqrt{85}}{85}; \cos\theta = \dfrac{2\sqrt{85}}{85};$

$\tan\theta = \dfrac{9}{2}; \cot\theta = \dfrac{2}{9};$

$\csc\theta = \dfrac{\sqrt{85}}{9}; \sec\theta = \dfrac{\sqrt{85}}{2}$

18. $P(-7, -5)$

$\sin\theta = -\dfrac{5\sqrt{74}}{74}; \cos\theta = -\dfrac{7\sqrt{74}}{74}$

$\tan\theta = \dfrac{5}{7}; \cot\theta = \dfrac{7}{5};$

$\csc\theta = -\dfrac{\sqrt{74}}{5}; \sec\theta = -\dfrac{\sqrt{74}}{7}$

Solve.

19. A circus performer trots her pony into the ring. The pony circles the ring 22 times as the performer flips and turns on the pony's back. At the end of the act, the pony exits on the side of the ring opposite its point of entry. Through how many degrees does the pony trot during the entire act?

$8100°$

Holt Algebra 2

Practice B

LESSON 13-3

The Unit Circle

Convert each measure from degrees to radians or from radians to degrees.

1. $\frac{5\pi}{12}$

$75°$

2. $215°$

$\frac{43\pi}{36}$ radians

3. $-\frac{29\pi}{18}$

$-290°$

4. $-180°$

$-\pi$ radians

5. $\frac{5\pi}{3}$

$300°$

6. $-\frac{7\pi}{6}$

$210°$

7. $400°$

$\frac{20\pi}{9}$ radians

8. $\frac{3\pi}{10}$

$54°$

9. $35°$

$\frac{7\pi}{36}$ radians

Use the unit circle to find the exact value of each trigonometric function.

10. $\cos\frac{2\pi}{3}$

$-\frac{1}{2}$

11. $\tan\frac{5\pi}{4}$

1

12. $\tan\frac{5\pi}{6}$

$-\frac{\sqrt{3}}{3}$

13. $\sin 315°$

$-\frac{\sqrt{2}}{2}$

14. $\cos 225°$

$-\frac{\sqrt{2}}{2}$

15. $\tan 60°$

$\sqrt{3}$

Use a reference angle to find the exact value of the sine, cosine, and tangent of each angle.

16. $150°$

$\frac{1}{2}; -\frac{\sqrt{3}}{2}; -\frac{\sqrt{3}}{3}$

17. $-225°$

$\frac{\sqrt{2}}{2}; -\frac{\sqrt{2}}{2}; -1$

18. $-300°$

$\frac{\sqrt{3}}{2}; \frac{1}{2}; \sqrt{3}$

19. $\frac{11\pi}{6}$

$-\frac{1}{2}; \frac{\sqrt{3}}{2}; -\frac{\sqrt{3}}{3}$

20. $-\frac{2\pi}{3}$

$-\frac{\sqrt{3}}{2}; -\frac{1}{2}; \sqrt{3}$

21. $\frac{5\pi}{4}$

$-\frac{\sqrt{2}}{2}; -\frac{\sqrt{2}}{2}; 1$

Solve.

22. San Antonio, Texas, is located about 30° north of the equator. If Earth's radius is about 3959 miles, approximately how many miles is San Antonio from the equator?

2073 mi

Holt Algebra 2

LESSON
13-4

Practice B
Inverses of Trigonometric Functions

Find all possible values of each expression.

1. $\sin^{-1}\left(-\frac{\sqrt{3}}{2}\right)$

$\dfrac{4\pi}{3}+2\pi n; \dfrac{5\pi}{3}+2\pi n$

2. $\cos^{-1}\left(-\frac{1}{2}\right)$

$\dfrac{2\pi}{3}+2\pi n; \dfrac{4\pi}{3}+2\pi n$

3. $\tan^{-1}0$

$0+2\pi n; \pi+2\pi n$

4. $\sin^{-1}\left(-\frac{\sqrt{2}}{2}\right)$

$\dfrac{5\pi}{4}+2\pi n; \dfrac{7\pi}{4}+2\pi n$

5. $\cos^{-1}\left(-\frac{\sqrt{2}}{2}\right)$

$\dfrac{3\pi}{4}+2\pi n; \dfrac{5\pi}{4}+2\pi n$

6. $\tan^{-1}\left(\frac{\sqrt{3}}{3}\right)$

$\dfrac{\pi}{6}+2\pi n; \dfrac{7\pi}{6}+2\pi n$

Evaluate each inverse trigonometric function. Give your answer in both radians and degrees.

7. $\text{Sin}^{-1}(-1)$

$\dfrac{3\pi}{2}; 270°$

8. $\text{Tan}^{-1}(-\sqrt{3})$

$\dfrac{5\pi}{3}; 300°$

9. $\text{Cos}^{-1}1$

$0; 0°$

10. $\text{Sin}^{-1}\left(\frac{\sqrt{3}}{2}\right)$

$\dfrac{\pi}{3}; 60°$

11. $\text{Tan}^{-1}\left(-\frac{\sqrt{3}}{3}\right)$

$\dfrac{11\pi}{6}; 330°$

12. $\text{Cos}^{-1}\left(\frac{\sqrt{2}}{2}\right)$

$\dfrac{\pi}{4}; 45°$

Solve each equation to the nearest tenth. Use the given restrictions.

13. $\sin\theta = 0.45$, for $0° < \theta < 90°$

$26.7°$

14. $\sin\theta = 0.801$, for $90° < \theta < 270°$

$233.2°$

15. $\tan\theta = 2.42$, for $180° < \theta < 360°$

$247.5°$

16. $\cos\theta = -0.334$, for $0° < \theta < 180°$

$109.5°$

17. $\cos\theta = -0.181$, for $180° < \theta < 360°$

$259.6°$

18. $\tan\theta = -10$, for $90° < \theta < 270°$

$95.7°$

Solve.

19. A 21-foot ladder is leaning against a building. The base of the ladder is 7 feet from the base of a building. To the nearest degree, what is the measure of the angle that the ladder makes with the ground?

$71°$

Holt Algebra 2

LESSON 13-5

Practice B
The Law of Sines

Find the area of each triangle. Round to the nearest tenth.

1.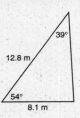

   39°

   12.8 m

   54°
   8.1 m

   41.9 m²

2.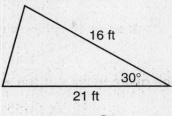

   16 ft

   30°

   21 ft

   84 ft²

3.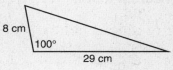

   8 cm

   100°

   29 cm

   114.2 cm²

Solve each triangle. Round to the nearest tenth.

4.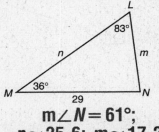

   L
   83°
   n   m
   36°
   M   29   N

   m∠N = 61°;
   n ≈ 25.6; m ≈ 17.2

5.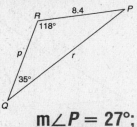

   R   8.4   P
   118°
   p   r
   35°
   Q

   m∠P = 27°;
   r ≈ 12.9; p ≈ 6.6

6.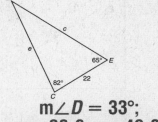

   D
   c
   e   65° E
   82°   22
   C

   m∠D = 33°;
   e ≈ 36.6; c ≈ 40.0

7.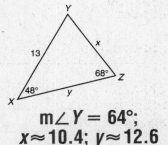

   Y
   13   x
   68° Z
   48°   y
   X

   m∠Y = 64°;
   x ≈ 10.4; y ≈ 12.6

8.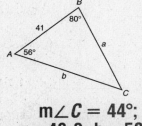

   B
   80°
   41
   A 56°   a
   b
   C

   m∠C = 44°;
   a ≈ 48.9; b ≈ 58.1

9.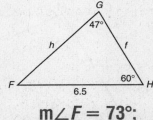

   G
   47°
   h   f
   60°
   F   6.5   H

   m∠F = 73°;
   h ≈ 7.7; f ≈ 8.5

An artist is designing triangular mosaic tiles. Determine the number of triangles he can form from the given side and angle measures. Then solve the triangles. Round to the nearest tenth.

10. $a = 8$ cm, $b = 10$ cm,
    $A = 60°$

    0 triangles

11. $a = 18$ cm, $b = 15$ cm,
    $A = 85°$

    1 triangle; c ≈ 11.3 cm;
    m∠B = 56°;
    m∠C = 39°

12. $a = 22$ cm, $b = 15$ cm,
    $A = 120°$

    1 triangle; c ≈ 10.3 cm;
    m∠B = 36°;
    m∠C = 24°

Solve.

13. Ann is creating a triangular frame. Two angles and the
    included side of the frame measure 64°, 58°, and 38 centimeters,
    respectively. What are the lengths of the other two sides
    of the frame to the nearest tenth of a centimeter?

    38.0 cm; 40.3 cm

Holt Algebra 2

LESSON 13-6

Practice B
The Law of Cosines

Use the given measurements to solve each triangle. Round to the nearest tenth.

1.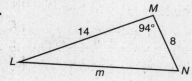

$m \approx 16.6$; $m\angle L \approx 28.7°$; $m\angle N \approx 57.3°$

2.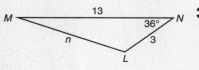

$n \approx 10.7$; $m\angle L \approx 134.5°$; $m\angle M \approx 9.5°$

3.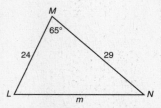

$m \approx 28.7$; $m\angle L \approx 65.9°$; $m\angle N \approx 49.1°$

4.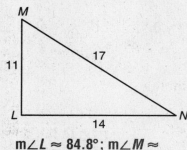

$m\angle L \approx 84.8°$; $m\angle M \approx 55.1°$; $m\angle N \approx 40.1°$

5.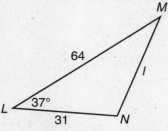

$l \approx 43.5$; $m\angle M \approx 25.4°$; $m\angle N \approx 117.6°$

6.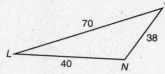

$m\angle L \approx 25.5°$; $m\angle M \approx 26.9°$; $m\angle N \approx 127.6°$

7.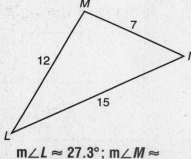

$m\angle L \approx 27.3°$; $m\angle M \approx 101°$; $m\angle N \approx 51.7°$

8.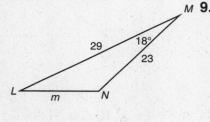

$m \approx 10.1$; $m\angle L \approx 44.7°$; $m\angle N \approx 117.3°$

9.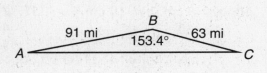

$m\angle L \approx 68°$; $m\angle M \approx 59.4°$; $m\angle N \approx 52.6°$

Solve.

10. A postal airplane leaves Island A and flies 91 miles to Island B. It drops off and picks up mail and flies 63 miles to Island C. After unloading and loading mail, the plane returns to Island A at an average rate of 300 miles per hour. How long does it take the pilot to travel from Island C to Island A?

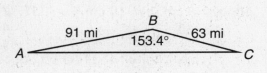

30 min

11. A statue is erected on a triangular marble base. The lengths of the sides of the triangle are 12 feet, 16 feet, and 18 feet. What is the area of the region at the base of the statue?

94.1 ft²

Holt Algebra 2

LESSON **Practice B**
14-1 *Graphs of Sine and Cosine*

Using $f(x) = \sin x$ or $g(x) = \cos x$ as a guide, graph each function. Identify the amplitude and period.

1. $b(x) = -5\sin\pi x$

Amplitude: 5; period: 2

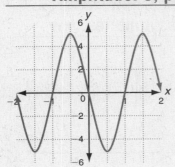

2. $k(x) = 3\cos 2\pi x$

Amplitude: 3; period: 1

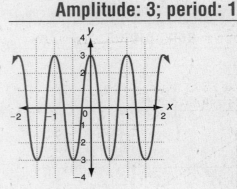

Using $f(x) = \sin x$ or $g(x) = \cos x$ as a guide, graph each function. Identify the x-intercepts and phase shift.

3. $h(x) = \sin\left(x + \frac{\pi}{4}\right)$

x-intercepts: $\frac{3\pi}{4}$, $\frac{7\pi}{4}$; phase shift: $\frac{\pi}{4}$ radians to the left

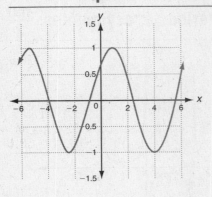

4. $h(x) = \cos\left(x - \frac{\pi}{4}\right)$

x-intercepts: $\frac{3\pi}{4}$, $\frac{7\pi}{4}$; phase shift: $\frac{\pi}{4}$ radians to the right

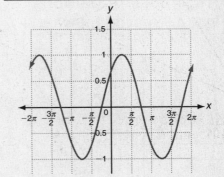

Solve.

5. a. Use a sine function to graph a sound wave with a period of 0.002 second and an amplitude of 2 centimeters.

b. Find the frequency in hertz for this sound wave.

500 Hz

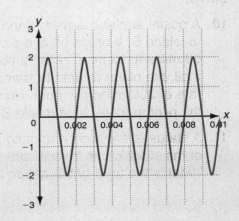

Holt Algebra 2

LESSON 14-2

Practice B
Graphs of Other Trigonometric Functions

Using $f(x) = \tan x$ and $f(x) = \cot x$ as a guide, graph each function.
Identify the period, x-intercepts, and asymptotes.

1. $g(x) = 2\tan\dfrac{\pi x}{2}$

Period: 2; x-intercepts: 2n;
asymptotes: 1 + 2n

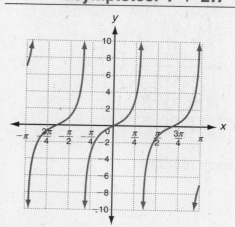

2. $t(x) = \dfrac{3}{4}\cot(x)$

Period: π; x-intercepts: $\dfrac{\pi}{2} + \pi n$;
asymptotes: πn

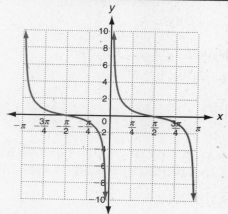

Using $f(x) = \cos x$ or $f(x) = \sin x$ as a guide, graph each function.
Identify the period and asymptotes.

3. $k(x) = \sec\dfrac{x}{2}$

Period: 4π; asymptotes: π + 2πn

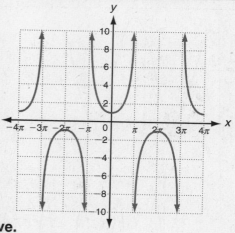

4. $q(x) = \dfrac{1}{2}\csc(2x)$

Period: π; asymptotes: $\dfrac{\pi n}{2}$

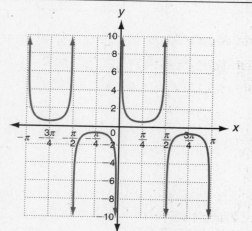

Solve.

5. The rotating light on a lighthouse is 400 feet from a cliff and completes one full rotation every 10 seconds. The equation representing the distance, a, in feet that the center of the circle of light is from the lighthouse is $a(t) = 400\sec\left(\dfrac{\pi t}{5}\right)$.

a. What is the period of $a(t)$?   **10**

b. Find the value of the function at $t = 10$.   **400 ft**

Holt Algebra 2

Practice B
LESSON 14-3

Fundamental Trigonometric Identities

Prove each trigonometric identity.

1. $\sin^2\theta + \sin^2\theta \cot^2\theta = 1$

$\sin^2\theta + \sin^2\theta \cot^2\theta = 1$

$\sin^2\theta\,(1 + \cot^2\theta) = 1$

$\sin^2\theta\,(\csc^2\theta) = 1$

$\sin^2\theta\left(\dfrac{1}{\sin^2\theta}\right) = 1$

$1 = 1$

2. $\cot^2\theta \cos^2\theta = \cot^2\theta - \cos^2\theta$

$\cot^2\theta \cos^2\theta = \cot^2\theta - \cos^2\theta =$

$\dfrac{\cos^2\theta}{\sin^2\theta} - \cos^2\theta = \dfrac{\cos^2\theta}{\sin^2\theta} - \dfrac{\sin^2\theta \cos^2\theta}{\sin^2\theta} =$

$\dfrac{\cos^2\theta - \sin^2\theta \cos^2\theta}{\sin^2\theta} = \dfrac{\cos^2\theta\,(1 - \sin^2\theta)}{\sin^2\theta} =$

$\left(\dfrac{\cos^2\theta}{\sin^2\theta}\right)(1 - \sin^2\theta) = \cot^2\theta\,(1 - \sin^2\theta) =$

$\cot^2\theta\,(\cos^2\theta + \sin^2\theta - \sin^2\theta) = \cot^2\theta \cos^2\theta$

3. $\tan^2\theta - \tan^2\theta \sin^2\theta = \sin^2\theta$

$\tan^2\theta - \tan^2\theta \sin^2\theta = \sin^2\theta$

$\tan^2\theta(1 - \sin^2\theta) = \sin^2\theta$

$\tan^2\theta\,(\cos^2\theta) = \sin^2\theta$

$\dfrac{\sin^2\theta}{\cos^2\theta}\,(\cos^2\theta) = \sin^2\theta$

$(\sin^2\theta)\,\dfrac{\cos^2\theta}{\cos^2\theta} = \sin^2\theta$

$\sin^2\theta = \sin^2\theta$

4. $\dfrac{\sin\theta + \cos\theta}{\sin\theta \cos\theta} = \sec\theta + \csc\theta$

$\dfrac{\sin\theta + \cos\theta}{\sin\theta \cos\theta} = \sec\theta + \csc\theta$

$\dfrac{\sin\theta}{\sin\theta \cos\theta} + \dfrac{\cos\theta}{\sin\theta \cos\theta} = \sec\theta + \csc\theta$

$\left(\dfrac{\sin\theta}{\sin\theta}\right)\dfrac{1}{\cos\theta} + \dfrac{1}{\sin\theta}\left(\dfrac{\cos\theta}{\cos\theta}\right) = \sec\theta + \csc\theta$

$(1)\dfrac{1}{\cos\theta} + \dfrac{1}{\sin\theta}(1) = \sec\theta + \csc\theta$

$\dfrac{1}{\cos\theta} + \dfrac{1}{\sin\theta} = \sec\theta + \csc\theta$

$\sec\theta + \csc\theta = \sec\theta + \csc\theta$

Rewrite each expression in terms of $\cos\theta$. Then simplify.

5. $2\sin\theta \cos\theta \cot\theta$

$2\sin\theta \cos\theta \cot\theta =$

$2\sin\theta \cos\theta \cdot \dfrac{\cos\theta}{\sin\theta} =$

$2\cos^2\theta$

6. $\dfrac{1 + \cot\theta}{\cot\theta(\sin\theta + \cos\theta)}$

$\dfrac{1 + \cot\theta}{\cot\theta(\sin\theta + \cos\theta)} =$

$\dfrac{1 + \dfrac{\cos\theta}{\sin\theta}}{\dfrac{\cos\theta}{\sin\theta}(\sin\theta + \cos\theta)} =$

$\dfrac{\dfrac{\sin\theta + \cos\theta}{\sin\theta}}{\dfrac{\cos\theta(\sin\theta + \cos\theta)}{\sin\theta}} =$

$\dfrac{\sin\theta + \cos\theta}{\cos\theta(\sin\theta + \cos\theta)} = \dfrac{1}{\cos\theta}$

7. $\cos^4\theta - \sin^4\theta + \sin^2\theta$

$\cos^4\theta - \sin^4\theta + \sin^2\theta$

$= (\cos^2\theta + \sin^2\theta)$

$(\cos^2\theta - \sin^2\theta) + \sin^2\theta$

$(1)(\cos^2\theta - \sin^2\theta) +$

$\sin^2\theta = \cos^2\theta$

Solve.

8. Use the equation $mg\sin\theta = \mu mg\cos\theta$ to determine the angle at which a waxed wood block on an inclined plane of wet snow begins to slide. Assume $\mu = 0.17$.

9.6°

Holt Algebra 2

Practice B
Sum and Difference Identities

Find the exact value of each expression.

1. $\cos 120°$

$$-\frac{1}{2}$$

2. $\sin 315°$

$$-\frac{\sqrt{2}}{2}$$

3. $\tan 255°$

$$2 + \sqrt{3}$$

4. $\tan \frac{7\pi}{6}$

$$\frac{\sqrt{3}}{3}$$

5. $\sin \frac{\pi}{12}$

$$\frac{\sqrt{6} - \sqrt{2}}{4}$$

6. $\cos \frac{3\pi}{4}$

$$-\frac{\sqrt{2}}{2}$$

Prove each identity.

7. $\sin\left(x - \frac{3\pi}{2}\right) = \cos x$

$$\sin\left(x - \frac{3\pi}{2}\right) = \cos x$$

$$\sin x \cos\left(\frac{3\pi}{2}\right) - \sin\left(\frac{3\pi}{2}\right)\cos x = \cos x$$

$$\sin x \cdot 0 - (-1)\cos x = \cos x$$

$$\cos x = \cos x$$

8. $\cos\left(x - \frac{\pi}{2}\right) = \sin x$

$$\cos\left(x - \frac{\pi}{2}\right) = \sin x$$

$$\cos x \cos\left(\frac{\pi}{2}\right) + \sin x \sin\left(\frac{\pi}{2}\right) = \sin x$$

$$\cos x \cdot 0 + \sin x \cdot 1 = \sin x$$

$$\sin x = \sin x$$

Find each value if $\cos A = \frac{12}{13}$ with $0° \leq A \leq 90°$ and if $\sin B = \frac{8}{17}$ with $90° \leq B \leq 180°$.

9. $\sin(A + B)$

$$\frac{21}{221}$$

10. $\cos(A + B)$

$$-\frac{220}{221}$$

11. $\tan(A + B)$

$$-\frac{21}{220}$$

12. $\sin(A - B)$

$$-\frac{171}{221}$$

13. $\cos(A - B)$

$$-\frac{140}{221}$$

14. $\tan(A - B)$

$$\frac{171}{140}$$

Solve.

15. Find the coordinates, to the nearest hundredth, of the vertices of triangle ABC with $A(1, 0)$, $B(10, 0)$, and $C(2, 6)$ after a 60° rotation about the origin.

a. Write the matrices for the rotation and for the points.

$$\begin{bmatrix} \cos 60° & -\sin 60° \\ \sin 60° & \cos 60° \end{bmatrix}\begin{bmatrix} 1 & 10 & 2 \\ 0 & 0 & 6 \end{bmatrix}$$

b. Find the matrix product.

$$\begin{bmatrix} 0.50 & 5.00 & -4.20 \\ 0.87 & 8.66 & 4.73 \end{bmatrix}$$

c. Write the coordinates.

$$A'(0.50, 0.87), \; B'(5, 8.66), \; C'(-4.2, 4.73)$$

16. A hill rises from the horizontal at a 15° angle. The road leading straight up the hill is 800 meters long. How much higher is the top of the hill than the base of the hill?

207 m

Holt Algebra 2

Name _____ Date _____ Class _____

Practice B
Double-Angle and Half-Angle Identities

Find sin 2θ, cos 2θ, and tan 2θ for each.

1. $\cos\theta = -\dfrac{12}{13}$ for $\pi < \theta < \dfrac{3\pi}{2}$

$$\frac{120}{169}, \frac{119}{169}, \frac{120}{119}$$

2. $\sin\theta = \dfrac{\sqrt{6}}{10}$ for $0 < \theta < \dfrac{\pi}{2}$

$$\frac{\sqrt{141}}{25}, \frac{22}{25}, \frac{\sqrt{141}}{22}$$

3. $\sin\theta = -\dfrac{2}{3}$ for $\dfrac{3\pi}{2} < \theta < 2\pi$

$$-\frac{4\sqrt{5}}{9}, \frac{1}{9}, -4\sqrt{5}$$

4. $\tan\theta = -\dfrac{5}{6}$ for $\dfrac{\pi}{2} < \theta < \pi$

$$-\frac{60}{61}, \frac{11}{61}, -\frac{60}{11}$$

Prove each identity.

5. $2\cos^2\theta = \cos 2\theta + 1$

$$2\cos^2\theta = \cos 2\theta + 1$$
$$2\cos^2\theta = \cos^2\theta - \sin^2\theta + 1$$
$$2\cos^2\theta = \cos^2\theta - (1 - \cos^2\theta) + 1$$
$$2\cos^2\theta = \cos^2\theta - 1 + \cos^2\theta + 1$$
$$2\cos^2\theta = 2\cos^2\theta$$

6. $\tan\theta = \dfrac{1 - \cos 2\theta}{\sin 2\theta}$

$$\tan\theta = \frac{1 - \cos 2\theta}{\sin 2\theta};$$
$$\tan\theta = \frac{1 - (1 - 2\sin^2\theta)}{2\sin\theta\cos\theta};$$
$$\tan\theta = \frac{1 - 1 + 2\sin^2\theta}{2\sin\theta\cos\theta};$$
$$\tan\theta = \frac{\sin\theta}{\cos\theta}; \tan\theta = \tan\theta$$

Use half-angle identities to find the exact value of each trigonometric expression.

7. $\tan 22.5°$

$$\sqrt{\frac{2 - \sqrt{2}}{2 + \sqrt{2}}}$$

8. $\cos\dfrac{7\pi}{12}$

$$-\frac{\sqrt{2 - \sqrt{3}}}{2}$$

9. $\sin\dfrac{11\pi}{12}$

$$\frac{\sqrt{2 - \sqrt{3}}}{2}$$

Find $\sin\dfrac{\theta}{2}$, $\cos\dfrac{\theta}{2}$, and $\tan\dfrac{\theta}{2}$ for each.

10. $\cos\theta = \dfrac{3}{5}$ and $270° < \theta < 360°$

$$\frac{\sqrt{5}}{5}, -\frac{2\sqrt{5}}{5}, -\frac{1}{2}$$

11. $\sin\theta = -\dfrac{\sqrt{5}}{3}$ and $180° < \theta < 270°$

$$\frac{\sqrt{30}}{6}, -\frac{\sqrt{6}}{6}, -\sqrt{5}$$

Solve.

12. A water-park slide covers 100 feet of horizontal space and is 36 feet high.

a. Write a trigonometric relation in terms of θ, the angle that the slide makes with the water surface.

$$\tan\theta = \frac{9}{25}$$

b. A new replacement slide will create an angle with the water surface that measures twice that of the original slide. The new slide will use the same horizontal space as the old slide. Write an expression that can be evaluated to find the height of the new slide.

$$\frac{100 \cdot 2\left(\frac{9}{25}\right)}{1 - \left(\frac{9}{25}\right)^2}$$

c. What is the height of the new slide to the nearest foot?

$$83 \text{ ft}$$

Holt Algebra

LESSON 14-6 Practice B
Solving Trigonometric Equations

Find all of the solutions of each equation.

1. $4\tan\theta = 5\tan\theta + \sqrt{3}$

$$120° + 180°n$$

2. $\sqrt{2} - 2\sin\theta = 0$

$$45° + 360°n, \ 135° + 360°n$$

3. $7\cos\theta - 1 = 9\cos\theta$

$$120° + 360°n, \ 240° + 360°n$$

4. $2\tan\theta + \sqrt{3} = 5\tan\theta$

$$30° + 180°n$$

Solve each equation for the given domain.

5. $\tan^2\theta - 2\tan\theta = -1$ for $0° \le \theta \le 360°$

$$45°, \ 225°$$

6. $2\sin^2\theta = 1$ for $0° \le \theta \le 360°$

$$45°, \ 135°, \ 225°, \ 315°$$

7. $4\cos^2\theta = 3\cos\theta$ for $90° \le \theta \le 180°$

$$90°$$

8. $2\sqrt{3}\cos^2\theta - \cos\theta = 2\sqrt{3}$ for $\pi \le \theta \le 2\pi$

$$\frac{7\pi}{6}$$

9. $3\tan^2\theta = 2\tan\theta + 1$ for $0° \le \theta \le 90°$

$$45°$$

10. $5\tan^2\theta = 2\tan\theta + 6$ for $180° \le \theta \le 360°$

$$232.7°, \ 317.6°$$

Use trigonometric identities to solve each equation for the given domain.

11. $\sin\theta + \sin 2\theta = 0$ for $0° \le \theta \le 360°$

$$0°, \ 120°, \ 180°, \ 240°, \ 360°$$

12. $\cos^2\theta = \sin^2\theta$ for $0° \le \theta \le 360°$

$$45°, \ 135°, \ 225°, \ 315°$$

13. $\dfrac{\cos 2\theta}{\sin^2\theta} = 0$ for $0° \le \theta \le 360°$

$$45°, \ 135°, \ 225°, \ 315°$$

14. $\cos 2\theta = \frac{1}{2}\sin\theta$ for $0° \le \theta \le 90°$

$$\approx 36.4°, \ \approx 57.5°$$

Solve.

15. The height of the water at a pier on a certain day can be modeled by $h(t) = 4.8\sin\frac{\pi}{6}(t + 3.5) + 9$, where h is the height in feet and t is the time in hours after midnight. When is the height of the water 6 feet?

$$\textbf{3:47 A.M., 7:13 A.M., 3:47 P.M., 7:13 P.M.}$$

Holt Algebra 2